Environmental Psychology

Annette Bolger

ABD PUBLISHERS
Jaipur ■ New Delhi

ABD PUBLISHERS

Regd. Off. "Bony Residency", Gate No. 2
Opp. Tilak Public School, Vishweswariya Nagar,
Gopalpura Road, Jaipur - 302018, India
Ph: 0141 - 2761280, Telefax: 2761381
Email: abdpublishersjpr@gmail.com

Delhi Office: 102, 1st Floor, Satyam House,
4327/3, Ansari Road, Darya Ganj,
New Delhi - 110002, India
Ph: 011-45652440

visit us: www.oxfordbookcompany.com

Distributed By
KEYA INTERNATIONAL PRESS LLP
E-mail: keyainternationalpressllp@gmail.com

Published in India, 2025

Price: ₹ 2195/-

ISBN: 9788183762427

Typeset by Shivangi Computer, Jaipur

Printed and bound in India by Thomson Press Ltd.,
New Delhi

Preface

The subject, psychology like other natural sciences has two aspects - pure and applied. The evolution of definition of the subject of psychology reflects the advent of the modern era of science and technology. Consequently, in the definition of psychology the word study was replaced by the word science. In this aspect of pure psychology it formulates broad principles, brings out theories and suggests techniques for the study of human behaviour which finds the practical shape in its applied aspect i.e., branches of applied psychology like occupational psychology, clinical psychology, crime psychology, industrial psychology, educational psychology etc.

Environmental psychology maintained its dual orientation to research and application and proceeded further toward integrating them. Researchers conducted hypothesis-testing studies in controlled setting and practitioners applied research findings in facility programming, design and post occupancy evaluation. The clearest integration appeared in environmental design research, natural setting to improve environmental design and to test scientific hypotheses.

Therefore, Psychology Environment is nothing but one of the branches of applied psychology. It is an attempt to apply the knowledge of psychology to the field of Environment. It consists of the application of the psychological principles and techniques to human behaviour in educational situations. In other words, Psychology Environmnet is a study of the experiences and behaviour of the learner in relation to educational environment.

Annette Bolger

Contents

Chapter 1

The Physical World

The *biosphere* is that part of Earth where life exists. In vertical dimension it extends from the deepest trenches in the ocean floor, more than 11,000 meters (36,000 feet) below sea level, to at least 10,000 meters (m) above sea level, where spores (reproductive cells) of bacteria and fungi can be found floating free in the atmosphere.

By far most living things — most of which depend directly or indirectly on the capture of solar energy by photosynthesis in plants and certain bacteria — exist in the narrower region extending from the limit of penetration of sunlight in the clearest oceans, less than 200 meters from the surface, to the highest value of the permanent snow line in tropical and subtropical mountain ranges — about 6000 meters, or 20,000 feet.

(Everest, the highest mountain, rises almost 8900 meters above sea level.) By any definition, the biosphere is as a mere film in thickness compared to the size of the ball of rock on which it sits — about like the skin of an apple, in fact. The radius of Earth is about 6370 kilometers (km), or 4000 miles (mi).

Of course, conditions within the thin envelope of the biosphere are influenced by physical processes taking place far outside it: by the energy emitted by the sun, 150 million kilometers away; by the tides originating in the relative motion and positions of Earth, sun, and moon (the distance from Earth to the moon averages about 380,000 km); by the presence of gases 20 to 400 km high in the atmosphere that screen out harmful components of incoming solar energy; by the constitution and structure of Earth's crust (to 40 km deep), which govern the availability of mineral nutrients at the surface and of metallic ores accessible to industrial civilization; by the behaviour of the solid but plastically flowing mantle (to 2900 km deep) on which the crustal plates "float" and move laterally; and by the motion of Earth's molten core

(2900 to 5200 km deep), which produces the magnetic field that protects the planet's surface from bombardment by energetic, electrically charged particles from space.

In horizontal extent the biosphere covers the globe, although in the hottest deserts and coldest polar regions — as at the highest elevations — usually only dormant spores can be found.

Earth's total surface area amounts to 510 million square kilometers (about 197 million mi), of which 71 per cent is ocean and 29 per cent land.

The mass of all living organisms on Earth amounts to about 5 trillion metric tons, three fourths of which consists of water.

Under the reasonable assumption that living matter has about the same density as water [1 gram per cubic centimeter (1g/cm)], this would mean that the living part of the biosphere was equivalent to a layer of material only I centimeter thick, covering the globe.

(The range concealed in this average is from 0.0002 g of living material for each square centimeter of surface in the open ocean to 15 g or more for each square centimeter of surface in a tropical forest.)

Table. Surface Areas of the Globe

		Percentage of category	Area (million km)
Total Earth Surface			510
Oceans			361
	Pacific Ocean	46	
	Atlantic Ocean	23	
	Indian Ocean	18	
	Arctic Ocean	4	
	Mean extent of sea ice	7	
Land			149
	Eurasia	36	
	Africa	20	
	North and Central America	16	
	South America	12	
	Antarctica	10	
	Oceania	6	
	Ice-covered land	10	
	Terrain more than 3000 meters high	5	
	Lakes and rivers	1	

Table. Masses of Constituents of the Physical World

Constituent	Mass (trillion MT)
Living organisms	8
Liquid fresh water, on surface	126
Atmosphere	5,140
Ice	30,000
Salts dissolved in oceans	49,000
Oceans	1,420,000
Earth's crust (average depth, 17 km)	24,000,000
Earth (total)	6,000,000,000

To study the processes that operate in any subdivision of the physical world — atmosphere, biosphere, Earth's crust — one must know something about energy: what it is, how it behaves, how it is measured. For whenever and wherever anything is happening, energy in some form is involved; it is in many respects the basic currency of the physical world.

An introduction to energy and the related concepts of work and power, along with the units in which these quantities can be measured. Some feeling for how much energy is stored in and flows between various parts of the physical world.

EARTH'S SOLID SURFACE

The outermost layer of Earth's solid surface is called the crust. It ranges in thickness from about 6 kilometers beneath the ocean floor to as much as 75 kilometers below the largest mountain ranges. In essence, the crust floats on the denser *mantle* beneath it. (The crust and mantle are differentiated by the different compositions and densities of the rock they comprise.) As with icebergs on the sea, the more crust extends above the surface (as in a mountain range), the more bulk is hidden below. This situation is made possible by the existence of a soft, yielding layer called the *asthenosphere* in the middle of the underlying mantle. This layer's strength is low because the rock is near its melting point. The combination of the crust and the hard upper layer of the mantle is called the *lithosphere,* a term sometimes also employed in a more general sense to mean the entire solid part of Earth. Below the mantle, between the depths of 2900 and 5200 kilometers, lies Earth's *molten outer core.* This core consists largely of liquid iron (with some nickel) at a temperature of perhaps 2500°C; its properties and motion produce Earth's magnetic field.

Table. Energy Flow and Storage in the Physical World

		Energy or power
Storage		**Trillion MJ**
	Energy released in a large volcanic eruption	100
	Chemical energy stored in all living organisms	30,000
	Energy released in a large earthquake	100,000
	Chemical energy stored in dead organic matter	100,000
	Heat stored in atmosphere	1,300,000
	Kinetic energy of Earth's rotation on its axis	250,000,000,000
Flows		**Million Mw**
	Tides	3
	Heat flow from Earth's interior	32
	Conversion of sunlight to chemical energy in photosynthesis	100
	Conversion of sunlight to energy of motion of atmosphere	1,000
	Sunlight striking top of atmosphere	172,000

Work, Energy and Power

Work is the application of a force through a distance. *Energy* is stored work. *Power* is the rate of flow of energy, or the rate at which work is done. All these concepts are more easily understood with the help of examples and some elaboration.

Work — force multiplied by distance — is done when a weight is lifted against the force of gravity (as with water carried upward in the atmosphere in the course of the hydrologic cycle), when mass is accelerated against the resistance of inertia (as with waves whipped up on the ocean by the wind) or when a body is pushed or pulled through a resisting medium (as with an aircraft moving through the atmosphere or a plow cutting through a field).

The presence of distance in the concept of work means that work is done only if there is motion — if we push on a stalled car and it doesn't budge, there is a force, but there is no work because there is no motion.

The foregoing are examples of *mechanical work* — work involving the bulk (or *macroscopic*) motion of agglomerations of molecules. There are also various forms of *microscopic* work, such as *chemical work* and *electrical work,* which involve forces and motions on the scale of individual molecules, atoms, and electrons.

To heat a substance is to do a form of microscopic work in which the individual molecules of the substance are made to move more rapidly about in all directions, without any bulk motion taking place. The demonstration that all these different manifestations of work are fundamentally the same can be found in treatises on physics and chemistry.

If work has many guises, so must energy, which is only *stored* work. Work stored as the motion of a macroscopic object (for example, a speeding automobile or the Earth spinning on its axis) is called *mechanical energy* or *kinetic energy*. The latter term may be applied as well to the energy of motion of microscopic objects (such as, molecules, electrons).

Work stored as the *disordered* motion of molecules — that is, rotation, vibration, and random linear motion not associated with bulk motion of the substance — is called *thermal energy* or *sensible heat* or (more commonly) just *heat*. Note that temperature and heat are not the same. Temperature is a measure of the intensity of the disordered motion of a typical molecule in a substance; the heat in a substance is the sum of the energies stored in the disordered motion of all its molecules.

Kinetic energy means something is happening; that is, the work is stored as motion. *Potential* or *latent* energy means something is "waiting" to happen. That is, the work is stored in the position or structure of objects that are subject to a force and a restraint; the force provides the potential for converting position or structure into kinetic energy, and the restraint is what keeps this from happening (at least temporarily).

Each kind of potential energy is associated with a specific kind of force. Gravitational potential energy (an avalanche waiting to fall) and electrical potential energy (oppositely charged clouds waiting for a lightning stroke to surge between them) are associated with forces that can act between objects at large distances.

Chemical potential energy (gasoline waiting to be burned, carbohydrate waiting to be metabolized) is associated with the forces that hold atoms together in molecules — that is, with chemical bonds. Nuclear potential energy is due to the forces that hold protons and neutrons together in the nucleus — the so-called strong force.

Latent heat of vaporization (water vapour waiting to condense into liquid, whereupon the latent heat will be converted to sensible heat) and latent heat of fusion (liquid waiting to freeze into a solid, with the same result) are associated with the electrical forces between molecules in liquids and solids.

The idea that potential energy is something "waiting" to happen needs only to be tempered by recognition that sometimes it can be a long wait — the chemical potential energy in a piece of coal buried in Earth's crust, for example, may already have waited a hundred million years.

Electromagnetic radiation is a form of energy that does not fall neatly into any of the categories we have mentioned so far. It is characterized not in terms of the motion or position or structure of objects but in terms of the motion of electric and magnetic forces. Light (visible electromagnetic radiation), radio waves, thermal (infrared) radiation, and X-rays are all closely related varieties of this particular form of energy.

Albert Einstein theorized, and many experiments have subsequently verified, that any change in the energy associated with an object (regardless of the form of the energy) is accompanied by a corresponding change in mass. In this sense, mass and energy are equivalent and interchangeable, the formal expression of equivalence being Einstein's famous formula $E = mc^2$.

(Here *E* denotes energy, *m* mass, and *c* the speed of light.) Because a small amount of mass is equivalent to a very large quantity of energy, a change in mass is only detectable when the change in energy is very large — as, for example, in nuclear explosions.

Different professions use a bewildering array of units for counting work and energy. The metric system is prevailing, but so gradually that the literature of energy and environmental sciences will be littered for years to come by a needless profusion of archaic units. Since work has the dimensions of force times distance, and all energy can be thought of as stored work, it should be apparent that a single unit will suffice for all forms of energy and work.

The most logical one is the *joule* (J), which is exactly the amount of work done in exerting the basic metric unit of force, 1 *newton* (N), over the basic metric unit of distance, 1 meter. We shall use the joule and its multiples, the kilojoule (1000 J, or kJ) and the megajoule (1,000,000 J, or MJ), throughout this book.

Our only exception to the use of the joule is a concession to the enormous inertia of custom in the field of nutrition, where we reluctantly employ the *kilocalorie* (kcal). A kilocalorie is approximately the amount of thermal energy needed to raise the temperature of 1 kg of water by 1 degree Celsius (1°C); this unit is often confusingly written as "calorie" in discussions of nutrition. Running the bodily machinery of an average adult human being uses about 2500 kcal — about 10,000 kJ — per day.)

Besides the erg (1 ten-millionth of a joule) and the calorie (1 thousandth of a kcal), the unit of energy most likely to be encountered by the reader elsewhere is the British thermal unit (Btu), which is approximately the amount of thermal energy needed to raise I pound of water by 1 degree Fahrenheit (1°F). A Btu is roughly a kilojoule.

In many applications one must consider not only *amounts* of energy but the *rate* at which energy flows or is used. The rate of energy flow or use is *power. Useful power* is the rate at which the flow of energy actually accomplishes work. The units for power are units of energy divided by units of time — for example, British thermal units per hour, kilocalories per minute, and joules per second. One joule per second is a *watt* (w).

A kilowatt (kw) is 1000 watts, and a megawatt (Mw) is 1,000,000 watts; these are the units we use for power in this book. These units are perfectly applicable to flows of nonelectric as well as electric energy, although we may be accustomed to them only in the context of electricity. Similarly, the kilowatt hour (kwh), denoting the amount of energy that flows in an hour if the rate (power) is 1 kilowatt, makes sense as a unit of energy outside the electrical context. A kilowatt hour is 3600 kilojoules.

For example, we can speak of an automobile using energy (in this case, chemical energy stored in gasoline) at a rate of 100 kilowatts (100 kilojoules of chemical energy per second). In an hour of steady driving at this rate of fuel consumption, the automobile uses 100 kJ/sec multiplied by 3600 sec (the number of seconds in an hour) or 360,000 kJ. The same quantity of energy used in a jumbo jetliner would produce a much larger power -say, 180,000 kw — for a much shorter time (2 sec).

Within the molten outer core is a *solid inner core,* also composed of iron and nickel, under enormous pressure.

Many of the characteristics of Earth's solid surface are the result of the operation of *tectonic processes*—the motion of great solid segments of the lithosphere, called *plates,* which slide about over the plastically flowing asthenosphere at a rate of a few centimeters per year. Operating over hundreds of millions of years, such motions have apparently produced displacements of thousands of kilometers.

The now widely accepted theory of continental drift holds that the present arrangement of the continents arose in this way, beginning with the breakup of the single supercontinent Pangaea about 200 million years ago. Some of the main tectonic processes, as they continue to work today. At *divergent plate boundaries* on the ocean floor, such as the East Pacific and Mid-Atlantic Ridges, adjacent plates move apart and new crust is created in the gap by *magma* molten rock which rises from below and then solidifies.

This phenomenon is called seafloor spreading. At *convergent plate boundaries*, as along the western edge of South America, one plate may be driven beneath the other into the asthenosphere. Heat generated by the friction in these *subduction zones* melts some of the crustal rock to produce magma, which rises to feed volcanic activity at the surface.

Deep-sea trenches, steep mountain ranges, and powerful earthquakes are other characteristics of these zones of violent collision between plates. At a third type of interface between plates, the plates slide past each other, moving parallel to the boundary. These *parallel-plate boundaries* are characterized by earthquakes with large surface displacements; the San Andreas Fault, which produced the great San Francisco earthquake of 1906, marks such a boundary.

Many other geophysical processes operate simultaneously with the tectonic motions described above to govern the shape and composition of Earth's crust. These processes include mountain-building by uplifting of the crust, the wearing-away of exposed rock surfaces by the actions of wind, rain, ice, and chemical processes (together these effects are called *weathering*), the transport of particles of rock and soil by water and wind (*erosion*), and the formation and transformation of new rocks from sedimentary material.

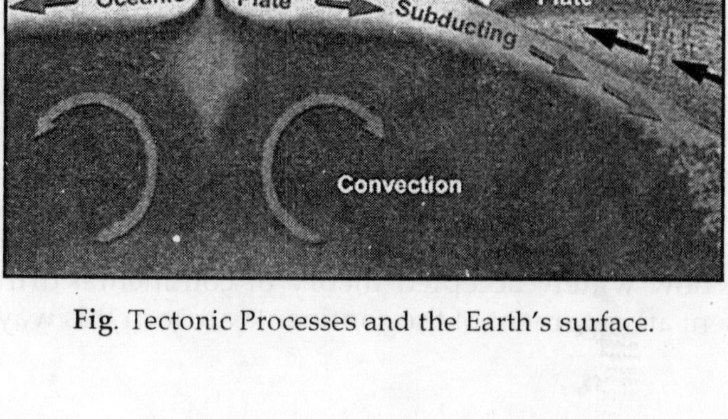

Fig. Tectonic Processes and the Earth's surface.

Rock that is exposed at the surface of the crust is gradually weathered away by physical and chemical processes. The resulting particles are some of the raw materials for new soil (the formation of which also requires the action of living organisms), and some of the chemicals liberated from the rock become available to the biosphere as nutrients. Although the rock particles may sometimes be carried uphill by wind and ice, the predominant motion is downhill with the flow of water.

Thus it happens over geologic spans of time (hundreds of thousands to millions of years) that large amounts of material are removed from the exposed rocky crust at high elevations and deposited on the lowlands and on the ocean floors. The accumulating weight of these sediments, consisting ultimately not only of rock fragments but also of dead plant and animal matter and chemicals precipitated out of seawater, contributes to sinking of the underlying crust and upper mantle.

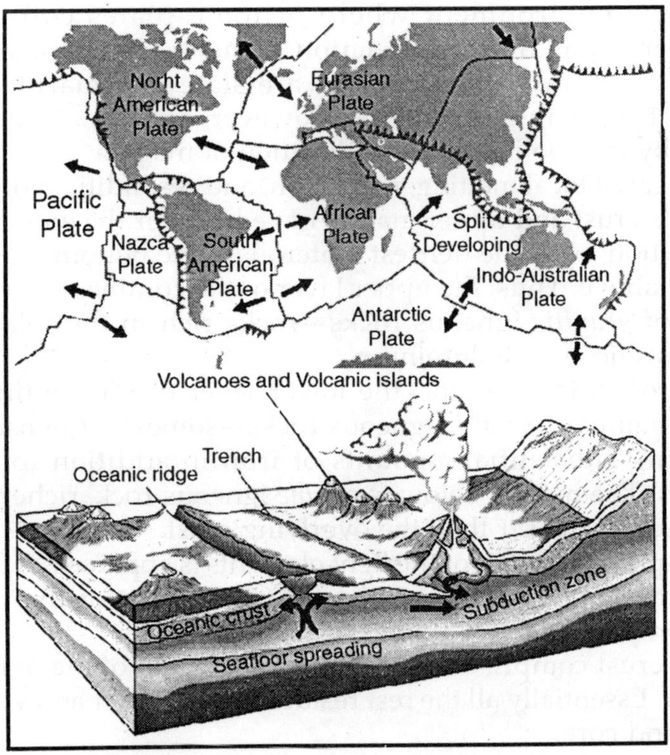

Fig. Tectonic Plates of the Lithosphere.

(Tectonic subsidence, the result of large-scale crustal motions, is more important in this sinking phenomenon than is the local accumulation of sediment weight, however.) Simultaneously, crustal rise, or uplifting, under the lightened regions restores some of the loss of elevation produced by weathering and erosion.

This process of sinking and uplifting is made possible by the capacity of the dense but soft (almost molten) asthenosphere to be deformed and, indeed, to flow.

The folding and buckling of Earth's crust, which has produced much of the varied topography is the combined result of the sinking-uplifting phenomenon just described and the continuous collision of the great lithospheric plates.

As layers of sediment become more deeply buried, they are subjected to temperatures and pressures high enough to initiate chemical and physical changes that transform the sediments into rock (called *sedimentary rocks*). Among the rocks formed in this way are shale, sandstone, limestone, and dolomite.

Under some conditions, such as the particularly energetic geological environment where tectonic plates collide, further transformations under the action of heat and pressure produce *metamorphic rocks*, among which are slate and marble. The most abundant rocks in Earth's crust, however, are *igneous rocks* — those formed by the cooling and solidification of magma.

Repeated local melting, migration, and resolidification of the rock in Earth's crust and upper mantle have led over the eons to a general stratification, with the densest material on the bottom and less dense material above. Thus, the upper layer of the continental crust consists largely of *granitic* igneous rocks—rocks rich in the relatively light elements silicon and aluminum.

The oceanic crust and the lower layer of the continental crust consist mainly of *basaltic* igneous rocks—somewhat denser material, containing substantial amounts of iron in addition to the lighter elements. The mantle below is *olivine* igneous rock, richer yet in iron and therefore denser than the overlying crust.

The predominance of the light elements is apparent: of the ten most abundant elements—accounting for 99 per cent of the mass of the crust—only iron has an atomic number above 25.

The crust comprises only about 0.4 per cent of the mass of Earth, however. Essentially all the rest resides in the denser and vastly thicker mantle and core.

The composition of the entire planet reflects the predominance of iron in those inner layers. Of interest is that carbon, the basic building block of living material, is not among the most abundant elements (it ranks fourteenth in crustal abundance, at 0.032 per cent).

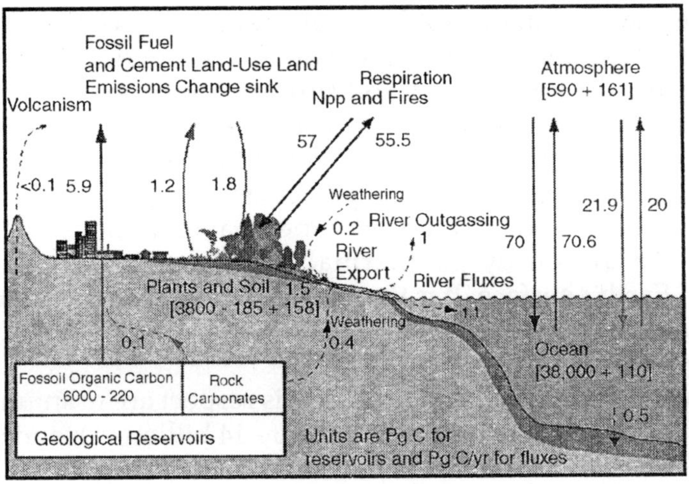

Fig. Geologic Cycles.

Table. Average Composition of Earth's Crust

Element	**Atomic number**	**Percentage by weight**
Oxygen	8	45.2
Silicon	14	27.2
Aluminum	13	8.0
Iron	26	5.8
Calcium	20	5.1
Magnesium	12	2.8
Sodium	11	2.3
Potassium	19	1.7
Titanium	22	0.9
Hydrogen	1	0.14

Table. Average Composition of Earth (Overall)

Element	**Atomic number**	**Percentage by weight**
Iron	26	34.6
Oxygen	8	29.5
Silicon	14	15.2
Magnesium	12	12.7
Nickel	28	2.4
Sulfur	16	1.9
Calcium	20	1.1
Aluminum	13	1.1
Sodium	11	0.57
Chromium	24	0.26

The energy that drives the great geological cycles has two distinct origins. Those parts of the cycles that take place on the surface—weathering, the formation of soil, erosion, the production of plant and animal matter that contributes to sediments—are powered by solar energy and its derivatives, wind and falling water.

The remaining geophysical processes (for example, the production and migration of magma and the inexorable motions of the tectonic plates) are driven by geothermal energy—heat that is produced beneath Earth's surface. It is thought that most of this heat results from the decay of radioactive isotopes that were already present when Earth was formed.

The most important isotopes in this respect are uranium-238 (half-life 4.5 billion yr), thorium-232 (half-life 14 billion yr), and potassium-40 (half-life 1.3 billion yr). Notwithstanding the rather low concentration of these isotopes in Earth's crust, the energy released by their continuing radioactive decay is enough to account approximately for the observed rate of heat flow to the surface. The very long half-lives of these isotopes guarantee, moreover, that this source of energy for geological change will have been diminished only slightly a billion years hence.

The processes of melting and resolidification, sinking and uplifting, the motion of tectonic plates of continental scale, the gouging and pushing of massive glaciers, and the different rates of weathering and erosion associated with different climates and different combinations of exposed rocks have combined to produce a tremendous variety of geological features.

The importance of these features to human beings is several fold. The landforms—plains, mountains, valleys, and so on—are one major determinant of the extent to which different parts of the planet's surface are habitable. The soils that have resulted from geological and biological processes over the millennia are another (and more limiting) Acterminant of how many people can be supported and where.

The zones of earthquakes and volcanism present serious environmental hazards to humans. And the distribution of fossil fuels and metals in scattered deposits far richer than the average crustal abundance is a geological phenomenon of enormous practical importance.

Although the processes that produced these features often act imperceptibly slowly in human terms, the temptation to consider the geological forces to be beyond human influence—or to take for granted their contributions to human well-being—must be resisted. Soil that has taken a thousand years to accumulate can be washed or blown away in a day through human carelessness; and there is evidence that the activities of human societies worldwide—cultivation, overgrazing, deforestation, construction—have doubled the prehistoric rate of sediment transport to the sea.

Earthquakes are widely feared and are called natural disasters, but the lack of foresight in planning and construction that has characterized the development of human settlements in active earthquake zones suggests that the consequences are due as much to human ignorance and irresponsibility as to nature's harshness.

There are circumstances, moreover, in which human activities actually *cause* earthquakes (injection of liquid wastes into rock formations, for example), and conceivably there may someday be technological means by which the frequency of strong earthquakes can be diminished.

Without concentrated deposits of mineral ores, industrial civilization as we know it could not have arisen; they represent a coincidental natural subsidy for society, provided by the work of natural energy flows over eons. The notion that technological civilization is now clever enough to do without this subsidy, once it is used up, by extracting needed materials from common rock is a dubious one.

THE HYDROSPHERE

Most forms of life on Earth require the simultaneous availability of mineral nutrients, certain gases, and water in liquid form. The boundaries of the biosphere—which are fuzzy, rather than sharp—can be defined as the places where the concentration of one or more of these essentials drops too low to sustain life. The principal reservoirs of available mineral nutrients are soil and sediment, the main reservoir of the needed gases is the atmosphere, and the primary supply of water is, of course, in the oceans.

Where they meet, these reservoirs intermingle to produce the most fertile parts of the biosphere: the upper layers of soil, where gases and moisture readily penetrate, and the shallower parts of the oceans, where nutrients from the land and the bottom mingle with dissolved gases and light that penetrates downward from the surface.

The oceans include not only some of the planet's most hospitable environments for life (and, almost surely, the environment where life began) but also make up by far the largest single habitat on Earth's surface. They cover almost 71 per cent of the planet and their volume is an almost incomprehensible 1.37 billion cubic kilometers (330 million mi).

The term *hydrosphere* refers not only to the oceans themselves, however, but also to the "extensions" of the oceans in other realms—the water vapour and water droplets in the atmosphere, the lakes and the rivers; the water in soil and in pockets deep in layers of rock; the water locked up in ice caps and glaciers.

The sections that follow here examine, first, some of the important characteristics of the oceans, then the behaviour of ice on Earth's surface, and, finally, the hydrologic cycle, which makes water so widely available even far from the seas.

The Oceans

More than 97 per cent of the water on or near the surface of Earth is in the oceans. This enormous reservoir is a brine (salts dissolved in water) of almost uniform composition. The concentration of the dissolved salts ranges from 3.45 per cent (by weight) to about 3.65 per cent, varying with depth and latitude.

The density of seawater varies between 1.026 and 1.030 grams per cubic centimeter, depending on depth and salinity, compared to 1.000 grams per cubic centimeter for fresh water at the reference temperature of 4°C (39°F).

An average cubic meter of seawater weighs 1027 kilograms of which about 36 kilograms is dissolved salts. Although most of this material is the familiar sodium chloride, more than half of all the known elements are present in seawater at trace concentrations or more.

Table. Water Storage in the Hydrosphere

Storage	Volume (1000 km)
Average in stream channels	1
Vapor and clouds in atmosphere	13
Soil water	67
Saline lakes and inland seas	104
Freshwater lakes	125
Groundwater (half less than 800 m below Earth's surface)	8,300
Ice caps and glaciers	29,200
Oceans	1,370,000
1 km = 264 billion gallons.	
Twenty percent of this total is in Lake Baikal in the Soviet Union.	
Source: Brian J. Skinner, Earth resources.	

Table. Composition of Seawater (excluding dissolved gases) (MT/km^3)

Elements at more than 1000 MT/km		Selected elements at less than 1000 MT/km	
H_2O	991,000,000	Lithium	175.0
Chlorine	19,600,000	Phospohorus	70.0
Sodium	10,900,000	Iodine	60.0
Magnesium	1,400,000	Molybdenum	10.0
Sulfur	920,000	Copper	3.0
Calcium	420,000	Uranium	3.0
Potassium	390,000	Nickel	2.0
Bromine	67,000	Cesium	0.4
Carbon	29,000	Silver	0.2
Strontium	8,300	Thorium	0.04
Boron	5,000	Lead	0.02
Silicon	3,100	Mercury	0.02
Fluorine	1,300	Gold	0.004

It is thought that this composition has remained essentially unchanged during most of geologic time. This would mean that the inflow of minerals reaching the oceans from rivers, from the atmosphere, and from undersea volcanoes has been roughly balanced by the outflow—namely, the incorporation of inorganic precipitates and dead organic matter into sediments on the ocean floor.

In a situation of *equilibrium* of this kind (with inflow balancing outflow it is easy to calculate the average time an atom of a given element spends in the ocean between entering it and leaving it. This is called the *residence time;* it is an important concept in the study of nutrient cycles and of pollution.

Flows, Stocks, and Equilibrium

The terms *mass balance, energy balance, input/output analysis,* and *balancing the books* all refer to fundamentally the same kind of calculation—one that finds extensive application in physics, chemistry, biology, and economics and in the many disciplines where these sciences are put to use. The basic idea is very simple: everything has to go somewhere, and it is possible and useful to keep track of where and how fast it goes.

A stock is a supply of something in a particular place—money in a savings account, water in a lake, a particular element in the ocean. The stocks can be measured in terms of value (dollars), volume (liters), mass (grams), energy (joules), number of molecules, or other units, but not time. Time appears, instead, in the complementary concept of *flows*, the inflow (or input) being the amount of commodity added to the stock per unit of time and the outflow (or output) being the amount of commodity removed from the stock per unit of time.

Thus, flows are measured in units like dollars per year, liters per minute, grams per day, or joules per second (watts). The sizes of the flows are indicated by the widths of the arrows. In a savings account, the inflow is deposits plus interest, the outflow is withdrawals, and the stock is the balance at any given time.

Clearly, if the inflow is greater than the outflow, the stock becomes larger as time passes; if outflow exceeds inflow, the stock shrinks. The change in the size of the stock in a given period is the difference between inflow and outflow, multiplied by the length of the period. In the event that the inflow and the outflow have exactly the same magnitude, the size of the stock remains constant.

This last situation, where inflow and outflow balance is called equilibrium, or, more specifically, *dynamic equilibrium* (something is flowing, but nothing is changing). The more restrictive case where nothing at all is happening—that is, no inflow, no outflow—is called *static equilibrium*.

In a state of equilibrium, there is not necessarily any relation between the size of the flows (*throughput*) and the size of the stock. For example, a small lake in equilibrium may be fed by a large river and drained by an equally large one (small stock, large throughput), or a large lake in equilibrium may be fed by a small river and drained by an equally small one (large stock, small throughput).

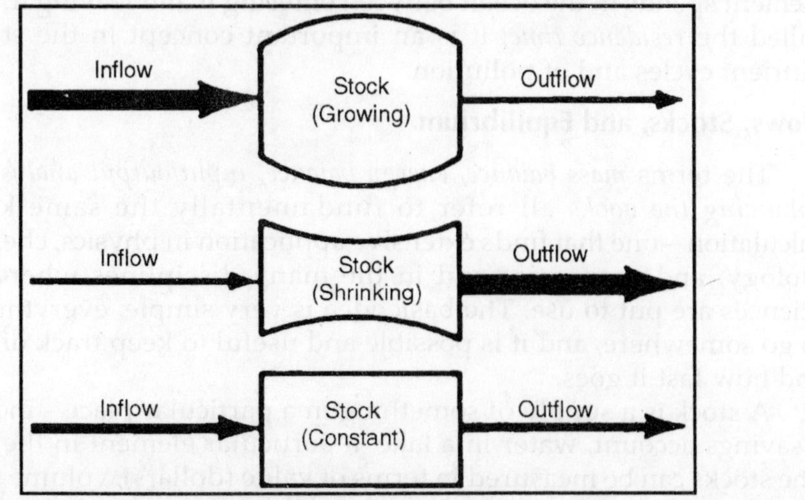

Fig. Relation between the Size of the Flows (*throughput*) and the Size of the Stock

If one divides the size of a stock in equilibrium by the size of the throughput, one obtains a very useful quantity—the *average residence time* (ô). In the example of the lakes, this is the average length of time a water molecule spends in the lake between entering and leaving. For a lake of 100 million cubic meters, fed and drained by two rivers with flows of 100 cubic meters per second each, the average residence time would be given by:

$$T = \frac{100{,}000{,}000\,m^3}{100 m^3/sec} = 1{,}000{,}000\,sec$$

which is about twelve days. A smaller volume and/or a larger throughput would produce a shorter residence time.

The concept of residence time is useful not only for describing geophysical processes but also for analyzing economic and biological ones. Economic "residence times" include replacement periods for capital and labour.

And, to cite an example from biology, if the stock is the world's human population, then the inflow is the rate at which people are born, the outflow is the rate at which people die, and the average residence time is the life expectancy. Clearly, a given population size could be maintained at equilibrium by conditions of high throughput (high birth rate, high death rate, short life expectancy) or by conditions of low throughput (low birth rate, low death rate, long life expectancy).

Table. Residence Times of Some Constituents of Seawater

Element	Residence time (million years)
Sodium	260
Magnesium	45
Calcium	8
Potassium	11
Silicon	0.01

The absolute quantities of materials dissolved in a cubic kilometre of seawater are quite large—175 metric tons of lithium, 3 metric tons of copper, 200 kilograms of silver, and 4 kilograms of gold, to mention some elements commonly regarded as scarce.

Multiplying such numbers by the total volume of the oceans gives very large numbers, indeed, and these staggering quantities, combined with the ready accessibility of the oceans, have stimulated much discussion of mining seawater for its riches. In mining, it is the *concentration* of the material that counts, however. To get the three tons of copper in a cubic kilometre of seawater, for instance, this desired material must somehow be separated from the billion metric tons of other elements mixed up with it, and this is not easy.

Probably a more important reason for looking into the details of the composition of the oceans is to evaluate the seriousness of various kinds of ocean pollution. If pollutants such as lead and mercury, for example, are added to the oceans in sufficient quantities to alter substantially the natural concentrations of those elements over large areas, one might suspect that significant biochemical consequences could result.

If, on the other hand, the discharge of an element into the ocean produces concentration changes small compared to natural variations in space and time of the concentration of this substance, little or no harm would be expected.

One cannot assume, of course, that substances added to the oceans are quickly diluted by this vast volume of water. How far? How deep? How fast? Are the questions, and the answers are found in the rather complicated patterns of horizontal circulation and vertical mixing in the oceans, as well as in the functioning of biological systems that may concentrate them.

Vertical mixing is rapid only near the surface. The turbulence (violent mixing motions) produced by wave action at the surface penetrates only to a depth of from 100 to 200 meters, and this defines the thickness of the layer within which most of the absorbed solar energy is distributed.

Below this warm, well-stirred surface layer is a transitional region called the *thermocline,* where the temperature drops rapidly. In this region there is usually less vertical motion than in the surface layer. Here heat penetrates partly by conduction (molecules passing on energy by jostling their neighbors) but mostly by convection (transport of energy by bulk motion of a warm medium) in large, slow eddies. There are two reasons for the relative lack of vertical motion in the thermocline:

- Motions originating at the surface have been damped out by friction before penetrating so deep
- The colder water near the bottom of the thermocline tends to be denser than the warmer water near the top of this layer, stifling thermal circulation.

In some circumstances, however, variations in salinity can influence density enough to produce a vertical circulation, despite the countervailing influence of the temperature profile. The bottom of the thermocline lies between 1000 and 1500 meters below the surface; from this level down, the temperature is nearly uniform and lies in the range from 0° to 5°C. (Seawater freezes at -2°C, or about 28.4°F.)

Like the thermocline, this deepest ocean layer is thermally

stratified, with the coldest water lying at the bottom. In the deep layer, moreover, salinity increases with depth; the saltier water is, the denser it is, so the salinity profile and the temperature profile both place the densest water at the bottom, inhibiting vertical mixing.

The stratification breaks down near the poles, where the cold layer extends all the way to the surface. That the surface waters in the Arctic and Antarctic oceans should be considered outcroppings of the deep layer that extends throughout the world ocean is suggested not only by patterns of temperature and salinity but also by the distribution of certain creatures.

The huge Greenland shark, for example, once thought to inhabit only Arctic waters, has been photographed three or four kilometers deep in waters off Baja California and in the Indian ocean.

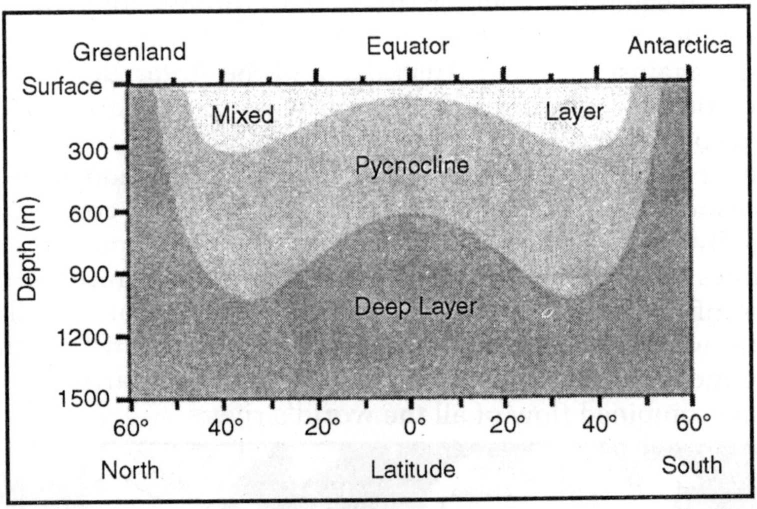

Fig. Vertical Structure of Oceans.

The stable stratification of the oceans also breaks down at scattered places and times far from the poles, as in upwellings in which winds push the surface water away from a steep continental slope and cold water rises from below to replace it (off the coast of Peru, for example) or when rapid cooling of surface water under unusual circumstances causes it to sink. The mean residence times for water in the various ocean layers illustrate the relative rarity in space and time of large vertical movements: a typical water molecule in the mixed layer may spend 10 years there, whereas one in the thermocline spends 500 years before reaching the deep layer, and a water molecule in the deep layer typically spends 2000 or 3000 years before reaching one of the upper layers.

Clearly, it must be assumed that most substances added to the oceans near the surface and dissolved there will remain near the surface for years, being diluted only by the small fraction of the ocean water that makes up the well mixed top layer (between 3 and 5 per cent).

Horizontal circulation in the oceans is considerably faster than vertical mixing. Water in the main currents, which generally involve only the mixed layers, typically moves at speeds of 1 kilometre per hour (km/hr), and occasionally up to 5 kilometers per hour.

Thus, an object or a substance being carried in the current might easily move 1000 kilometers in a month and cross an ocean in, six months to a year.

The principal features are the circular movements, called *gyres*, centered in the subtropical latitudes (25° to 30° north and south of the equator), an equatorial countercurrent (most prominent in the Pacific) that provides a return pathway for water that would otherwise be piling up against Asia and Australia, and the Antarctic circumpolar current, flowing uninterrupted from west to east around the far southern part of the globe.

These currents are produced by a complex interaction of the effects of the winds, Earth's rotation, and the placement of continents and islands. By moving enormous quantities of water-sometimes warm, sometimes cold—from one region to another, the ocean currents exert a major influence on climate, a subject taken up in more detail later.

The two mightiest currents on Earth, the Antarctic circumpolar current and the main branch of the Gulf Stream, each carries some fifty times the combined flow of all the world's rivers.

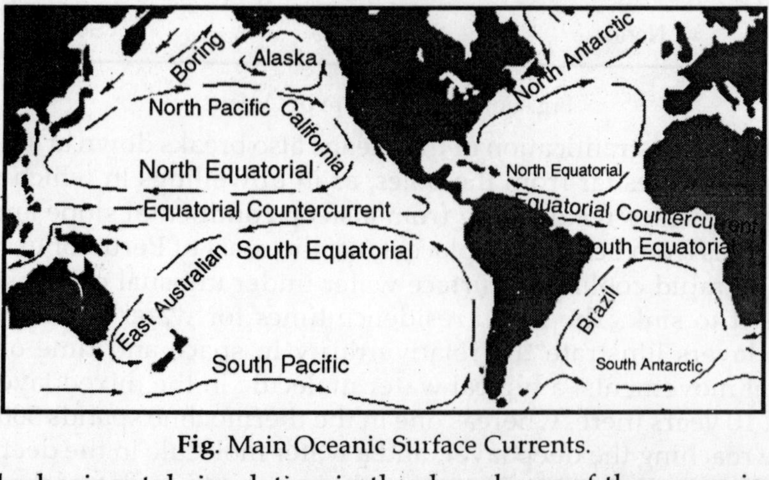

Fig. Main Oceanic Surface Currents.

The horizontal circulation in the deep layer of the ocean is much less thoroughly mapped than that of the surface layer and has been widely supposed to be much less vigorous. Typical speeds in these deep currents have been thought to be on the order of 0.1 kilometre per hour or less.

An increasing number of direct measurements of deep ocean currents now suggest a much more vigorous deep-ocean circulation, however, involving powerful eddies 100 kilometers or more in horizontal extent, containing currents of 0.5 to 1 kilometre per hour. One would expect the general flow to be from the poles toward the equator, inasmuch as some cold water enters the deep layer from above at the poles and some rises in upwellings closer to the equator.

But the actual situation is made quite complicated by Earth's rotation, by the irregular distribution of landmasses, and by the complex topography of the ocean floor.

It has been true historically and is still true today that the usefulness of the oceans to civilization and, in turn, civilization's impact on the oceans have been greatest in the shallower waters at the edges of the continents. This is so partly for simple reasons of accessibility, and partly because of the particular fertility of the near-shore waters and the richness of the underlying sediment in minerals of economic interest.

The term *continental shelf* refers to that part of the near-shore underwater topography that is actually an extension of flat lands on a continent itself.

Although the outer edge of a continental shelf is often defined as the line along which the depth of the water reaches 200 meters, a less arbitrary boundary is the point where there is a marked increase in the downward slope of the bottom.

The steeply sloping region just beyond this boundary is called the *continental slope*. The "foothills" leading from the ocean floor to the seaward edge of the continental slope are called the *continental rise*.

Using the definition of continental shelf just given, it has been estimated that continental shelves underlie 7.5 per cent of the area of the oceans (an area equal, however, to 18 per cent of Earth's land area). These shelves vary in width from essentially nothing to 1500 kilometers, and their seaward edges vary in depth from 20 to 550 meters (the average depth at the edge is 133 meters).

The circulation patterns in shallow, continental-shelf waters are complex, and the residence times of dissolved substances over the shelf can be surprisingly long — as much as several years to migrate from the coastline to the outer edge of a wide shelf like that off the east coast of the United States.

Chapter 2

Atmospheric Sound

Like titles in this one, too, lacks a lot of making clear all it is meant to imply, that is, the effect of the state and condition of the atmosphere on the distribution of sound, and the kinds of sounds that are induced by this or that sort of weather, or are associated with it.

In short, "weather music" is here meant to cover exactly the same field as does "meteorological acoustics" in its widest sense, but in a less formidable manner than that which would be necessary if one stopped to prove anew the correctness of every assertion by the strait-jacket logic of mathematics. Nevertheless, every statement made has had either just such proof, or the verification of experiment, or both.

THE SPEED OF SOUND

Observation shows us that while the noise made by a hammer, for instance, that we ourselves are wielding seems to be simultaneous with the blow, that made by a hammer half a mile away lags far behind the stroke. Naturally, this arouses the interest of any healthy boy, or of any one else, for that matter, whose curiosity bump has not atrophied. And so measurements were made long ago to determine how fast sound travels in the air.

But the discovery that this was about 1,100 feet per second, and the same in every direction, when the air is still, did not remain an isolated and sterile fact, for immediately the inquisitive boy wanted to know why sound had this velocity in air, whether this velocity is the same for all sounds, whether it is the same at different temperatures, different pressures and different humidities, and whether it is the same for all gases, and a thousand other "whethers," some of which even yet are not fully answered, nor, of course, will that be the end of the matter when they are.

Pretty soon it became evident from straight-forward, though none too easy, logic that the velocity of sound through a gas should vary directly as the square root of its "spring," or pressure it exerts per unit area, and inversely as the square root of its density, or mass per unit volume. This relation was first established theoretically by Newton. No fault could be found with his reasoning, and yet the calculated velocity of sound in the air was some 20 per cent less than the observed velocity.

Of course Nature was right and even Newton wrong, but for 100 years no one could find the "nigger in the wood pile."

And there he remained hidden until the keen vision of Laplace located him among the density changes of the air from compression to rarefaction, incident to the passage of a sound wave, which changes, as he proved, are much less than those assumed by Newton and his followers—less because the pressure changes are accompanied by temperature changes which Newton had overlooked. This slight but important correction brought the theory into perfect agreement with observation and experiment.

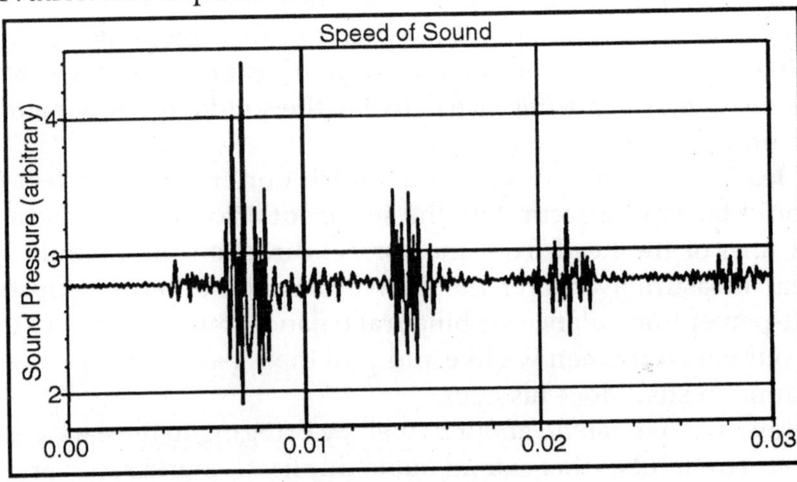

Fig. Speed of Sound

There is nothing in this theory to indicate that the velocity of sound should have one value for noise and another for music, nor that it should vary with either loudness or pitch, and experiment shows that in fact it is the same for all sounds however greatly they may differ in origin, character, or quality, so long as they are true sounds and not powerful explosion waves.

Neither is this velocity affected by any change in the pressure alone, since such change simultaneously increases, or decreases, as the ease may be, in exactly the same proportion, both the "spring" (pressure) and the density of the air, that is, both the numerator and the denominator of the fraction (thus leaving its value unaltered) whose square root is proportional to the velocity in question. On the other hand this velocity is altered by a change in the temperature of the air, because change of temperature, while leaving the pressure the same, so alters the density that it is all the time inversely proportional to the absolute temperature.

Hence, other things being equal, the velocity of sound in the air is directly proportional to the square root of the absolute temperature. Again, the velocity of sound in the atmosphere varies slightly with the humidity, chiefly, but not wholly, because the greater the humidity the less the density of the air and therefore the faster the travel of sound through it.

SOUND POINTING

It is a common observation that any animal with two good ears can locate pretty well the direction from which a sound is coming, and it is our experience that we hear a sound loudest and best when its source is directly in front of us, or when the air waves reach both ears simultaneously and in the same condition, or phase. Indeed, we are very sensitive to any difference between the sound in one ear and that in the other.

Clearly, therefore, if we could stretch our ears out ten feet or so we could face quite accurately the source of a sound, because a very slight turn of the head from the correct direction then would bring one ear measurably nearer the source than the other and upset the sound-perception balance, or binaural balance. But we can't do that—push our ears out when we like and pull them back as we please, like the common snail does his eyes.

However, we can do about as well by bringing to the ears we have suitable, equal tubes that extend out to duplicate sound receivers, such as long conical horns so mounted that their axes are parallel, and the whole revolvable as a unit in the proper manner.

When one plane in while the source of the sound lies is known, the horizontal, for instance, as when the source is on or near the ground in a level region, two listening tubes, lying in that plane, are sufficient; but when it is up in the sky somewhere, an airplane for example, four receivers, one at each corner of a square, are necessary.

If two of these are in a vertical plane they will give the elevation of the source while the other pair will give its azimuth, and thus the two pairs together its exact direction, or direction of the position it (the plane say) had when the sound started, except, in all cases, for such corrections as may be necessary owing, especially, to winds and differences in temperature.

SOUND RANGING

Where the source of a sound is on a level plane, to take a simple case, and the air is still, we can not only point towards it, but even determine just where it is by getting its directions from two known places, provided these directions are not the same. Evidently it is where these two directions cross each other.

This is simple in principle, but in the case of a single pulse, as from the firing of a cannon, it is more practical and convenient to have at each of three points, some distance apart, a suitable receiver that electrically records the arrival of the sound on a common-to-all timing device, and from the data thus obtained find the location of the cannon, say, as follows: Mark, on suitable paper, the three observing stations, correctly located with reference to each other.

About *B* and *C*, the second and third stations reached by the sound, describe circles whose radii, in proper scale, are equal, each to the distance the sound traveled in the interval between the time it arrived at the first station *A* and the station in question.

Then by trial find the circle whose circumference passes through *A* and is tangent to the circles about *B* and *C*. The centre of this last circle is the position of the cannon we are trying to locate.

Corrections have to be made for wind direction and velocity and for any other disturbing factors there may be at the time and place. Of course, too, the details of the procedure may differ from those here outlined, but the principle will remain the same.

DOPPLER EFFECT

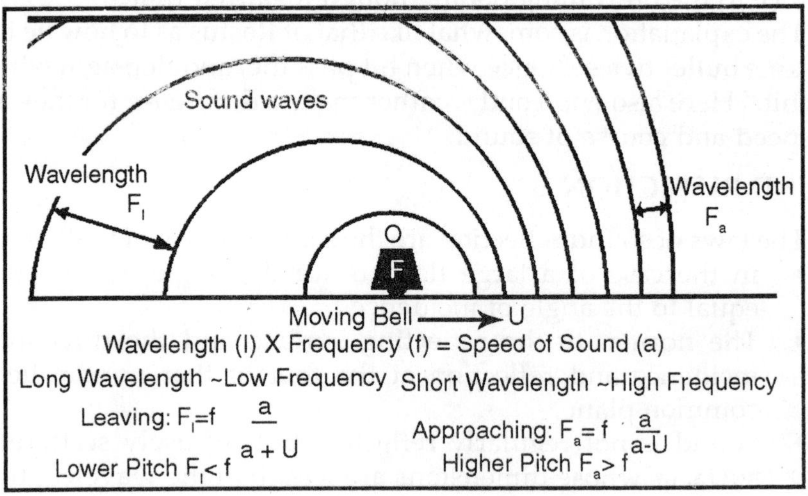

Fig. Doppler Effect on Sound Wave

Since sound travels at a moderate speed through the air, it follows that our rapid approach towards the source of a musical note, or its rapid approach towards us, causes a greater number of sound waves to reach our ears per second than would otherwise, and therefore the note to have a correspondingly higher pitch.

Conversely, when the source is receding from us, or we from it, the pitch is lowered.

This Doppler effect, as it is called after the name of the one who first explained it, also, like many others, is affected by the condition of the, especially as to its temperature and humidity, for changes in these particulars alter the velocity of sound and therefore the extent to which its pitch is raised or lowered by a given speed of approach or recession, respectively, between source and ear.

Thus, this effect is greater in winter by 5 per cent, roughly, in middle latitudes, than in summer because the velocity of sound in air is directly proportional to the square root of the absolute temperature. The humidity effect is very much less.

SOUNDS OUT OF ORDER

Many a soldier has been surprised at seeing a sudden flash from an enemy's cannon, then hearing, first the whine of the projectile passing by, and finally the boom of the gun. This hearing of the last-made sound first, and the first last, is at once explained by the fact that the velocity of a cannon ball over much of its course is greatly in excess of that of sound through the air, so much greater, indeed, that often the distance to the projectile at its nearest approach is far less than the distance to the on-coming boom, though it started first.

The explanation is somewhat like that of Rastus as to how he heard the same bullet twice: "Once when hit pass me, and den agin when ah pass hit." Here also wind and weather make a difference for they affect the speed and course of sound.

SOUND REFLECTION

The laws of sound reflection are the same as those of light, namely:

- In the case of a large flat surface the angle of reflection is equal to the angle of incidence.
- The normal to the reflecting surface and the directions of incidence and reflection at the foot of that normal lie in a common plane.
- Sound is not regularly reflected, but diffusely scattered, by surfaces whose dimensions are less than the wavelength.

Evidently, then, a sound incident onto a surface of considerable extent of a liquid, or solid that, like glass, is free from energy-absorbing pores, is nearly all reflected, owing to the great difference between the densities of the two media. Sound also is reflected at an abrupt, or nearly abrupt, interface between two layers or masses of air of different densities, but is not reflected where the change in density from the one to the other is gradual over a considerable number of wave lengths.

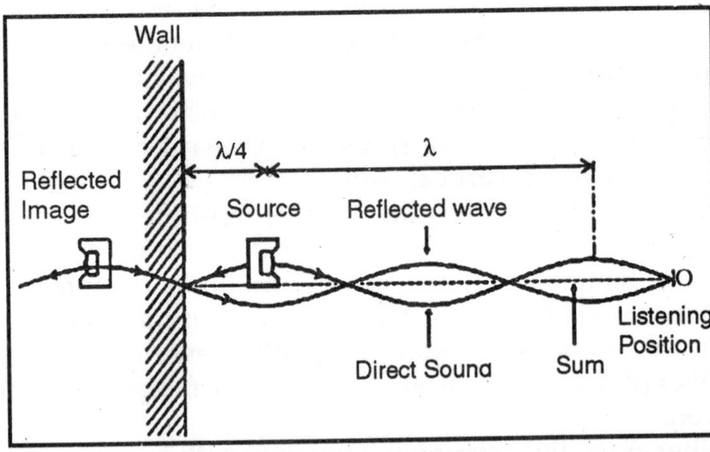

Fig. Reflection of Sound

The ratio of the intensity of the reflected sound, however, in free air to that of the incident is very small for normal incidence, but increases rapidly as the angle of incidence grows larger, to total reflection, on going from the rarer to the denser air, when the sine of that angle is equal to the square root of the ratio of the two densities.

Thus, if the densities of the two adjacent layers or masses of air differ by one per cent, or are to each other as 99 to 100, as sometimes can be the case, the intensity of the normally transmitted sound is about 160,000 times greater than that of the reflected, but only 100 times greater when the angle of incidence is 80°. Finally, at 841/4° all the sound is reflected and none transmitted.

Clearly, therefore, the familiar old assertion that the rumbling of thunder is owing to reflections by clouds is just a guess, and a wrong one at that.

It is pertinent here to ask what becomes of all the sound that goes skyward if a gradual change in density, such as takes place in the atmosphere with increase of height, causes no reflection. The answer, of course, is that as the density becomes less and the free paths of the air molecules greater, more and more of the mechanical energy of the sound is converted into heat until no sound is left.

This has never been offered, and is not here offered, as an

important contributing factor to the high temperature, if indeed it is really high, of the upper welkin.

ECHO

Much of interest might be said about the reflection of sound back to its origin, or more generally, its strong reflection in any direction; but while this is a striking outdoor phenomenon it is not enough affected by weather conditions to justify any detailed discussion of it in this connection. It will be interesting however to mention certain more or less distinct classes of these reflections or echoes. One of the most common is the discrete single echo, heard but once, and best produced by a solitary vertical wall.

Another common type is the discrete multiple echo, the sort, for instance, one hears over and over again every time a single pulse, from the blow of a hammer perhaps, passes him on its rebounds back and forth between parallel vertical walls. A more common type than either of these is the overlapping multiple echo, or reverberation, produced by a number of unequally distant reflecting surfaces, such as the walls of a canyon.

Another exceedigly common, though usually inconspicuous, kind is the diffuse echo, produced by objects which, like the leaves of a tree, are so small that they do not regularly reflect, but merely scatter the incident sound. Akin to this is the well known, but seldom experienced, harmonie echo, in which, owing to the proper sizes of the reflecting surfaces the overtones of a complex sound are more fully reflected than the fundamental, resulting in an echo of higher pitch than the initial sound. Finally we the musical echo, that affords the small boy, and the grown-up too, so much fun.

A fine example of it is the song of a flight of many steps, such as are in front of certain public buildings, in response to a loud clap of the hands, or other single sound pulse. The steps being equally spaced produce just so many echos at equal time intervals and therefore a musical note whose pitch is fixed by the width of the treads. Another good example, the one the boy especially likes, is the whistle of the picket fence in response to the crack of a whip—interesting and surpriseful when we know the tricks.

SOUND REFRACTION

Since the velocity of sound in still air increases as the density decreases—varies inversely as the square root of the density—it follows that the course of sound abruptly changes its direction wherever incident slantingly onto the interface between two masses of air of different temperatures or different water vapour contents. The amount of this sudden change of course is given by the simple relation: The sine of the angle of refraction divided by the sine of the angle of incidence is equal to the square root of the ratio of the density of the heavy air to that of the lighter.

When the change of density is gradual, and not abrupt, the direction of the course of sound through still air, oblique to the planes of equal density, also changes gradually, and in such manner as to make the path concave towards the denser air. In short, the laws of sound refraction are exactly the same as the well-known laws of light refraction.

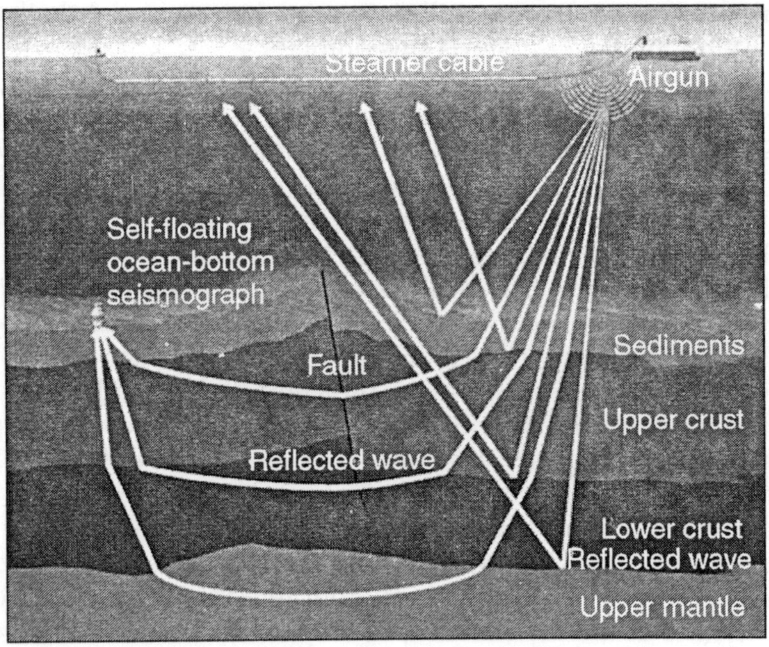

Fig. Refraction of Sound

SOUND DRIFT

Obviously drifting air, or a wind, carries along its state or condition—its temperature, humidity and even any wave pulse that may be in it, that is, sound is transported by wind, fastest in the direction the wind is fastest, and irregularly where the wind is irregular. Wind, therefore, produces very important effects on the velocity, spread and final disposition of sound. The effects, however, differ widely from each other and therefore must be considered under separate heads.

DISSIPATION OF SOUND

Common experience shows that in still air sound travels out from its source with equal velocity in every direction, and a little reasoning adds that when this source is over a considerable expanse of still water, for example, or level prairie, the primary sound is supplemented by its reflection that appears to have come simultaneously from the image of the source beneath the horizontal surface. Thus the firing of one gun may seem like the firing of two in quick succession.

A little further reasoning tells us that since the temperature of the air normally decreases with increase of height therefore the velocity of the sound also usually must decrease with gain of elevation. Hence, as a rule, each sound front in still air has the shape of a slightly flattened dome. Wind, however distorts this dome.

A steady wind leans it to leeward, and, if moderately strong, even lifts its windward side up above the surface so that there the sound may pass quite unnoticed above one's head though it still may be loud at double the distance n the opposite direction. Furthermore, when the distribution of temperature is irregular and the wind full of turbulence, as it normally is, then our sound dome is not only leaned over in the direction the wind is blowing but also distorted into numerous bulges and dents of various sizes and shapes; and the sound, traveling normally, as it must, to this irregular surface quickly loses unity of action, crisscrosses in every direction, and speedily fades away. Under such conditions, and they represent the usual state of the atmosphere, the intensity of sound often decreases far more rapidly than one would expect from the law of inverse squares.

The rate at which a spreading sound is lost or dissipated is affected also by humidity as Knudsen has shown, in the sense that the greater the amount of water vapour present, from zero to a few per cent relative humidity, the decidedly more rapid the fading of sound, and beyond that the considerably less rapid.

This surprising phenomenon, which becomes the more pronounced with the increase both of temperature and of pitch, has been explained by Kneser as owing to the finite time required for the vibrational and rotational energies of the disturbed molecules to come into equilibrium. Furthermore, this action of water vapour is through the oxygen of the air only, and not the nitrogen.

SPREADING OF SOUND

For the reasons just given, namely, irregular distribution of temperature and turbulence, leading each to an uneven wave front, sound often is quickly lost. When, however, the air is very still and much colder at and near the surface than it is a hundred feet or so above, as it is in the morning before sunup after a calm and clear winter night, sound can be heard to a surprising distance, especially over level land or a frozen lake.

This is because all the sound that leaves the source within a horizontal circular wedge whose angular thickness at the source is any thing up to 15° or possibly 20°, in extreme cases, is kept within a horizontal disk by reflection at the surface and progressive refraction above, which refraction is owing to increase of temperature with increase of height to the top of the cold or inversion layer.

In short, the cold layer of air, from the surface up to the level at which temperature ceases to increase with height, is a speaking disk, free from convection and wind turbulence, in which the intensity of sound is inversely proportional more nearly to the first power of the distance than to its square. Under such weather conditions a long, straight canyon acts somewhat as a speaking tube, and conveys sound correspondingly far.

If the lower air, instead of getting rapidly warmer with increase of height and thus forming a speaking disk, as explained, should get rapidly cooler, as it sometimes does over barren soil on a calm summer day, sound would not carry far along the surface, but, because moving fastest in the lower, warmer air, would lift from the surface by refraction, and pass over head at no great distance from the source. Excessive humidity confined to near the surface produces the same result to a greater or less degree.

This explains why, on such occasions, one may not hear a gun he sees fired less, sometimes, than two hundred yards away, or a field cannon over a distance as little as half a mile. Indeed there normally is a tendency for sounds made near the surface of the earth to lift, since usually the temperature of the air decreases with increase of height immediately from the surface.

This explains the long known fact that the higher bells are above the surface the greater the distance they are heard. There is a good reason, then, for putting a bell in a high steeple, even when that steeple is some distance from any structure that might interfere with the free passage of the sound waves.

If the composition and temperature of the air were constant from the surface up to a considerable height, any sound made in it would spread out spherically, but this condition practically never occurs.

SOUND MIRAGE

When there is no wind and the temperature drops rapidly with increase of height through 20 to 50 feet, or so, and then much less rapidly, or not at all, for some distance beyond, sound produced a short distance above the boundary between these layers produces some surprising effects, surprising because seldom noticed, and seldom noticed partly because the requisite state of the air in a strong degree is not very common, and partly because of our habitual inattention to such things.

The effect is an acoustical mirage. The sound wave in the lower stratum travels faster and faster as it approaches the surface, and finally a portion of it bends back up, as though reflected by a horizontal surface. A little farther on the sound is not heard at all in the lower warm layer.

Just above the boundary between the two strata, at the proper distance from the source, one should hear the same pulse twice in quick succession, or, if the sound is a continuous note, he should hear two identical notes, separated by a phase difference that depends on one's position.

One of these sounds seems to come from its actual source; the other from somewhere below that position. Furthermore, this "mirrored" sound is upside down in the sense that if two distinguishably different sounds originate one a little above the other, A above B, say, then in the "mirrored" sound B will seem to be above A.

Still higher in the upper stratum a single sound will be heard from a single source. In brief, all the acoustical effects produced under these conditions are the same as those in the well-known inferior optical mirage and are similarly explained.

Also the superior optical mirage and all its complexities have their acoustical duplicates. Indeed the very conditions that cause the one cause also the other. Thus, if, over a level region, there is no wind, and if the temperature of the air is constant through the first 20 to 30 feet above the surface, then increases appreciably with elevation through the next 20 to 30 feet to a uniform temperature for some distance beyond, queer and conspicuous optical effects may be observed, and a careful examination will reveal exactly similar sound phenomena.

Four listeners, one above the other, the first near the ground, the second in the upper portion of the lowest air stratum, the third in the middle stratum and the fourth in the upper stratum, would not all agree about a gun that was fired a short distance away.

The man on the ground and the highest of the four would agree with each other, and both be right, as to the location of the gun and that there was but one gun. The man in the middle air stratum would also say that there was one gun, but he would locate it much too high.

The man in the upper portion of the lower stratum, however, would differ from all the others and insist that there were three guns, one above the other, fired in rapid succession.

Such sound phenomena as these are not only interesting but sometimes of unexpected practical importance. For example, a fog horn, or other warning signal, may not be heard directly at all, while its reflection is, and the pilot guided thereby not to safety but to destruction. Certain actual disasters are believed to have been so caused.

Sound Before Rain

From ancient times it has been said that the carry of sound often is especially good for a few hours before the onset of a general rain, but the reasons why this is so are seldom presented because, in general, those who believe it can't explain it and those who can explain it don't believe it.

Well, it is true and there are several factors that combine to make it so. The sky is likely to be cloudy, and thus the sound-dissipating turbulence incident to strong surface heating, common to clear days, absent.

Also the vapour content of the air increases before a general rain, and this increases its carry of sound. Again, at that time the wind normally is from a warmer towards a cooler region, a course that sometimes results in making the air at very moderate heights nearly to quite as warm as, or even warmer than, the surface air, a condition that holds sound down near the ground and permits it to spread comparatively far before it becomes inaudible.

SOUNDS BETWEEN HIGH AND LOW PLACES

Another bit of ancient knowledge is the fact that a person on a mountain peak, or even in a high steeple, and best of all, as we now know, in a balloon, finds it far more difficult to get people on the plain below to hear him than he does to hear them.

The reason or reasons for this are not as clear and convincing as one would like, but one contributing cause is the fact that there is no sea of noises about the balloonist that can drown out sounds from beneath, while on the ground there commonly is a continuous, if highly varied, babel through which only a relatively loud noise can arrest attention.

Another factor, though of small importance, tending to the same result, is the decrease in density of the air with increase of height, since to halloo equally loud necessitates a greater effort on the part of the one above than the one below. Finally, the man up in the air gets both the direct and the ground-reflected voice of the one at the surface, each without disturbances by echo or otherwise from round about, while such echoes as the one below hears of the voice above come in irregular volume from various directions and merge with the innumerable distracting noises that normally keep all but the louder sounds beyond the threshold of one's attention.

ZONES OF SILENCE

Since World War a good deal of attention has been given to the spread of air-waves, both audible and inaudible, from the seat of great explosions, or even the firing of the largest cannon.

The essential features are:

- The noise is clearly heard all about in every direction for a distance of 30 miles, roughly, then for another 30 miles or so it is not heard, and after that again heard, often for a distance as great as, or even greater than, the width of the zone or ring of silence.
- The time of transit to the more distant listening posts is decidedly greater than would be required by sound coming directly near the surface.
- The sound front as it reaches the surface beyond the zone of silence is so inclined as to indicate that the source is at some considerable height.

These observations force the inference that the sound waves have traveled to high levels where they were reflected or refracted in such manner as to reach the earth again. A plausible explanation of this phenomenon is that it is owing to sound refraction, though this in turn requires the assumption that the air at 25 to 35 miles, or more, above the surface of the earth is fully as warm as, if not even much warmer than, the air at the surface. What this temperature really is no one knows.

A wide range of values has been computed, but none observed at greater heights than 18 to 20 miles, where it is around -60° F. in middle to high latitudes, to -90°F. in tropical regions.

THUNDER

Thunder is caused by an electric explosion of the air along the lightning path. The term "explosion" seems appropriate owing to the extremely sudden and violent expansion of the highly agitated and disintegrated air along this path, a path carrying a current of around 6000 amperes per square inch cross section.

The duration of thunder over level regions obviously is the time it takes sound to travel the greatest, minus the least, distance to the lightning that produced it. In mountainous regions the duration of a peal of thunder usually is more or less increased by reflection. The familiar booming and rumbling of thunder are owing, mainly, to irregularities in the path of discharge, a boom coming from a section of the path of appreciable length all parts of which are nearly equally distant from the observer.

Thunder impresses us as being about the loudest of all noises, yet it seldom is heard more than 12 to 15 miles, while a cannon of moderate size may be heard three or four times as far. There are two important differences in favour of the cannon. In the first place the cannon's noise is concentrated mainly in a single pulse while that of the thunder is spread out through a large volume. Indeed if all the noise of a loud peal of thunder were concentrated like that of a cannon it surely would "raise the roof" as the saying is.

Again, a cannon often is fired early in the morning when the air is still and the surface layer decidedly colder than that a short distance, 100 to 200 feet, above, and therefore its sound carrying capacity at its best, while thunder, on the other hand, starts in, and travels through, air so divided up by strong winds and eddies as to be a very poor transmitter of sound of any kind. In this way, too, the advantage is all in favour of the cannon.

Doubtless nearly every one hears thunder as a prolonged rumbling noise, fading away from an initial crash or boom to a vanishing murmur. But whoever pays close attention may sometimes, when the lightning is close by, hear the thunder sing. This is caused by a series of lightning discharges, the kind that produce flickering, along the same path and at so nearly equal time intervals that their successive thunder pulses jointly produce a musical note.

Often, too, a sound as of the tearing of paper, or the frying of bacon, some say, is heard close by an object for a second or two before it is struck. This seems to be the result of rapidly growing preliminary or brush discharges.

BRONTIDES

Brontides, or whatever else from their many local names one may call them, are muffled grumbles, like distant thunder of uncertain direction. They are best heard during the hush of a calm night or sultry afternoon, and since they occur chiefly in seismic regions it is believed that they are owing to earthquake tremors too feeble to be felt or even instrumentally recorded.

HUMMING OF WIRES

Every one is familiar with the hum of the telegraph wire, and knows too that it is loudest when the temperature is low and a moderate, but not too stiff, breeze is blowing across it. We used to be jokingly told, when credulous children, that this humming was caused by the messages in transit. Indeed that is one explanation of the phenomenon that has been proposed in all seriousness, and it is every whit as good, however absurd, as several others that have appeared where ordinarily one would not look for nonsense.

In reality it is produced by a vibrating eddy in the wake of the wire, or, more exactly, a succession of eddies starting alternately from first one side and then the other of the wire, which throw it into a corresponding vibration that is transmitted to the poles which, in turn, act as sounding boards.

When the wire, or other obstruction, has a natural or free vibration period, or periods (it may swing in segments) it commonly holds one or another particular pitch, as illustrated by the aeolian harp, through a considerable range of wind velocity, and if it changes at all does so abruptly from note to note among fundamental and harmonics.

This singing is loudest in winter, with equal wind velocity, because the wires then are stretched tightest and therefore transmit their vibrations in greatest force to the sounding boards—the poles. How these eddies behind the wire are started and maintained is a fascinating story, but one much too long to tell here—and not every detail of it is known either.

If the obstacle, a slender, cylindrical rod, for instance, is inelastic, a wind sets up eddies in its wake just the same, that produce a note whose pitch, inversely proportional to the diameter of the rod, increases directly with the speed of the air past it.

The note produced by a single such sound may be inaudible, and yet a multitude of them acting together and producing, as they do, a blended note of mean pitch, may produce a sound of considerable volume, because the average intensity of the resultant is the sum of the intensities of its many components.

DIRGE OF AN OLD PINE TREE

When one's mood is just right and he is all alone in a lonesome place nothing is more solemnly sweet than the dirge of an old pine tree. Under such a spell we are not apt to think of the physics involved, but if and whenever we do, then or later, the explanation is simple and obvious.

Each of the myriads of needles produces a faint aeolian tone as the wind blows across it, and the blend of the whole, with the variations in loudness and pitch that come with the ceaseless changes of the wind make up the dirge without end.

MURMUR OF THE FOREST

From the single pine tree it is easy to pass to a grove or a whole forest, where we find the effect to be the same in kind but greater in volume. Nor is it necessary that the trees be pines with long needles, that wail in high pitched sibilant cries. Oaks and maples, and all others, when their leaves are gone, gruffly grumble at the winter's blast, and the whole forest murmurs together like a vast multitude of living creatures.

Here the harp strings are the millions of twigs of many sizes, each feebly mumbling the one note it can, and must, as determined by its diameter and the speed of the wind. Though the voice of no twig alone is audible the joint effect of all the forest is a murmur that often can be heard a mile away, and sometimes much farther.

ROAR OF THE MOUNTAIN

Certain forest-clad mountains, when the trees are bare and the wind strong across their crests, produce muffled roars in the leeward valleys, as of distant mighty cataracts. These roars, however uncanny or even ominous they may seem (and they do impress those who have never heard them before) are only the united voice of myriads upon myriads of twigs each speaking in inaudible whispers—a voice aroused by the wind, and by it also roughly focused or concentrated onto the valley.

This focusing which greatly accentuates the effect is owing to the general trend of the wind down the mountain together with its increase of velocity with increase of height which bends the sound waves forward into a concentrating shape. And the people that dwell in the valley to the north (in the middle latitudes of the northern hemisphere) of such a mountain say, and say truly, that its roaring warns of an impending storm.

Nor can we claim the recognition of this warning to be any proof of the superiority of modern observation over ancient, for way back in biblical times Elijah told Ahab to hurry and eat and get down from Mount Carmel"for there is a sound of abundance of rain." And this sound it would seem could have been nothing other than the murmur of the forest mingled, perhaps, with the roar of the mountain.

"And it came to pass in the mean while, that the heaven was black with clouds and wind, and there was a great rain. " Elijah lays no claim to originality in his weather observation, and indeed it is highly probable that he got it from his elders as they did from theirs generation after generation back to that historyless past in which so much that is practical today had its origin.

ROAR OF THE TORNADO

The smallest yet most violent of all storms, the tornado, is as noisy as destructive, coming as it does with howls and shrieks in the sky and a roar and rumble on the ground as of many freight trains. The latter part of its bluster, the roar and rumble, is owing, in part at least, to the wreckage of everything encountered by the frenzied fury, even to the ripping up of sod and soil.

But whence the wild howls and shrieks from above, and even before the surface is touched? Seemingly they can have but one origin, namely, the vibrations of unstable shreds and tongues of air as they madly encircle and climb the whirling vortex, whose wall is surprisingly thin and semirigid, stressed by a spin of bullet velocity.

Howling of the Wind

The best place and time to hear the howling of the wind ("wild wailing wind of misfortune and sorrow") at its loudest, and to feel its spell to the fullest, is in an exposed country house on a bleak, winter night when the wind is blowing a gusty gale. And to heighten the effect one would better be quite alone and think of ghosts, ghouls, were wolves, demons and devils. That helps more than the physics of the phenomena to make the impression lasting! However, the cause really is mechanical and objective. The wind as it sweeps over the gables and past the chimneys is set into vibrating sheets and periodic eddies. It is they that do the howling and nothing uncanny after us from the underworld. The sounds are far louder inside the house than out, and while, as stated, we know their fundamental cause no one yet has solved the urbulence problem involved in all of its details. That is a demon too, of a different kind.

Boom of Lakes

Another uncanny sound, occasionally heard in lake regions when the weather is very cold, is the loud boom caused by the snapping and tearing of long rifts in thick ice as the result of a heavy strain induced by the temperature contrasts, or otherwise.

Bursting of Trees

Still another startling sound of weather origin, especially startling to one alone at night in a forest, the only place it is likely to be heard, and there but shortly after the onset of an exceptionally cold wave, is the crack of a pistol where there is no pistol and none to fire it.

This unexpected noise-without warning that may make the timid "jump out of his boots," is caused by the sudden splitting of the outer shell of some tree, strained beyond its strength by shrinking tighter and tighter, under the chill of the frigid air, about the relatively uncooled inner wood.

Shrieking and Howling of the Devil's Fiddle

Another sort of weather music, weird, uncanny, startling, is the shrieking and howling of the devil's fiddle. Here the music, always weird, is most effective when the setting and the circumstances insure acute and apprehensive attention: Place, deep in a vast forest roamed by wild beasts and said to be haunted; time, long after dark; circumstances, alone and a storm brewing.

Then as we quicken our steps there comes to our keenly attentive ears a faint howl from some ravenous beast far away, too far, we thank our stars, to be of immediate danger. Soon there is another howl, louder and closer than the first; others and still others, fast coming towards us; and now, as the tree tops are furiously shaken, a curdling shriek and yell from straight overhead. Right then and there, unless we have jumped to this music before, we leave our whole skin behind we.

Of these fiddles there are many sizes and grades. One of the most effective is a tall, dead tree lodged on falling in the high fork of a nearby companion whose swaying in the winds has worn both, at their place of contact, down to the hard and dry wood beneath. Thus worn, the two trees move back and forth on each other as swayed by the wind; not smoothly, but in myriads of jerks, much as the resined bow moves across the strings of a violin—jerks that knock the air into corresponding waves of sound, barely audible when the wind is light, to fearsome shrieks when it is high and gusty.

Such is the devil's fiddle, not a Stradivarius, though having some features in common with it, but under the right conditions just as effective in arousing our feelings.

The Cracking of Houses

The way for a small boy to know that there are spooks about is to lie awake in a frame country house during the wee hours of a calm and bitterly cold night, and listen to the snap, snap, bang of the joists, rafters and other parts as the stress of the cold jerks their joints into new adjustments.

He listens for the next bang and the next, every one of which comes unexpectedly and so mysteriously that after a while he gets jumpy too and fears the hobgoblins are about to get him—just an effect of the weather on the house.

Clatter of Hail

It is said that sometimes hailstones are heard bumping each other in mid air. Possibly so, but obviously that does not compare to the clatter they make on a tin roof, for instance, nor to the crashes they cause when falling on the glass cover of a greenhouse.

Hail, especially large hail when falling in quantity can do great damage and always makes a lot of fuss.

Rattle of Sleet

If it doesn't rattle-rattle on the windowpane—it isn't sleet, at least not as officially used by the U. S. Weather Bureau, and as sung by the poets.

These ice shot, or frozen raindrops, sometimes are driven along with considerable speed by the wind, and at such times, if falling fast, and the wind is gusty, as it often is, irregularly rattle on one's windows like separate handsfull of rice thrown at them.

Patter of Rain

Even a heavy rain of large drops makes a weather noise, a distinctive sort of patter, particularly when falling onto a body of water, or any hard flat surface, such as a paved street.

Every one is familiar with this patter of the rains, but whoever tries to explain it in all its details probably will come to the wise conclusion that the less said about it the better.

DETONATION OF METEORITES

The noise produced by a bursting meteorite is much like that which a cannon shell would cause if exploded at the same place. It does not seem to call for any special remarks, except, perhaps, to note that it is a far louder noise than the mere pulling in two of the meteorite by ordinary laboratory means probably would produce. Possibly some one may be able to explain why—if it really is.

When the electric strain is very great between the earth and air, all sorts of tall conductors, well connected to the earth —well grounded, we say—give off fitful streams of electric discharges from their topmost points. These streams consist, essentially, of innumerable tiny sparks each producing a correspondingly minute "explosion" of the air through which it passes. And as these discharges are of unequal volume, and of irregular occurrence the result is only a sizzling noise that generally is not heard more than a few feet away. Occasionally, however, the sparks grow larger and louder, and often then terminate in, or with, a lightning flash.

Swish of the Aurora

Many observers insist that they have heard the swish of the aurora as it swept past them. Others, with years of polar experience, say with equal emphasis that they never have heard any sound which they could attribute to the aurora. Finally, a few able scientists tell us that they too have heard it, but that it was something else.

They found the swishes to be timed in tune with their own breathing and to occur only when the temperature of the air was very low. From these facts they infer that the aurora's swish is nothing other than a faint noise caused by the sudden freezing of the moisture in the observer's breath. How that can cause a sound is not explained—another chance for some one with an inquisitive urge.

Four-Horse Wagon

When the temperature is way below freezing and there are a few inches of snow on the ground the steel-tired four-horse wagon is as happy as a lark, if we can judge by the beautiful way it sings as it rolls along. Children love to hear it and never hesitate to ask daddy a question that he cannot answer:

"What makes the wagon sing?" It isn't always musical, not even when there is plenty of snow on the ground, for when the temperature is but little below freezing the snow merely packs to streaks of ice and the rolling wheels are dumb. When, however, the temperature is very low the snow snaps and breaks like glass, and the wheels, leaving trails of powdered crystals behind them, sing their loudest and best.

But even then the automobile has no voice and the sleigh is silent, or nearly so, while beneath our heels as we trudge along there is crunching and nothing more. Presumably, therefore, it is elastic vibrations of the wheels, caused by the sudden slips and slides of the snow when too cold to fuse, that make the highpitched sweet music of the wagon. But how does the pitch vary with size of wheel, with its speed and with the load on the wagon? Who will find the answers?

Chapter 3

Atmospheric Optics

It may be insisted, with some show of reason, that optics is entirely an in-door matter, made up of beautiful parlor experiments—aesthetic physics, wholly unassociated with anything so commonplace as the weather. But many optical phenomena are seen only outdoors and some are so dependent on the state of the weather that they never appear except when that is just right.

Clearly, then, these are weather matters and no discussion of meteorology is complete, in the sense of covering the whole subject, without at least some account of them—some account of the things in the sky we misjudge owing to the deception of perspective; something about how the curved and even crooked courses of light through the air cause us to see things where they are not, and even things that are not there to see; something about those superglories of the heavens, the rainbow, the halo and the iridescent, or mother of-pearl, cloud; and something, too, about how the twilight sky can run the gamut of colors from red where the sun went down, or is soon to rise, through orange, yellow and green to blue, and even purple afar off.

Many of these accounts are best told in the concise and exact language of the mathematician, but with our kind permission the king's English will here be used—it reads better.

PERSPECTIVE-DECEPTIVE APPEARANCE

One of the most frequent of the weather's optical deceptions generally is called: "The sun drawing water." It is nothing but sunshine and shade in a normally dusty atmosphere—sunshine through clear spaces between broken patches of cloud, or notches in a distant horizon, and shades by the intervening cloud fragments, or mountain peaks.

Each sunlit shaft is made white with the light scattered by the dust it contains, while the shaded spaces, though equally dusty, are comparatively dark, scattering little light as they receive but little. So simple, the explanation, as to be all but obvious, and yet there are those, lots of them, who insist that these shafts are great pipes through which the sun pumps up water for the next rain. How this absurd idea ever got started is neither obvious nor a matter of record. Perhaps it first was just a dose of intellectual paregoric, unkindly given to an inquisitive child who never fully recovered from its effects.

The puzzling thing, a real puzzle, about these "crepuscular rays," as they properly are called, is the fact that they all seem to come from a common point just back of the clouds, or just beyond the horizon, as the case may be, in the direction of the sun, and to spread out from that point like the spokes of a wheel, or ribs of a giant fan, when the sun that casts them is ninety odd millions of miles away, and they therefore actually as parallel to each other as parallel can be.

Why this illusion, why do those rays we know to be parallel seem to run in every direction of the compass? Owing to perspective, we usually say, which, being interpreted, means that they seem to fan out because they seem to behave that way—a vicious circle, from which the only escape is by way of an adequate understanding of perspective.

Now the meaning of this term is: The appearance to the eye of objects in respect to their positions only, or, in other words, to what their projections on an intervening screen perpendicular to the line of sight would look like—a gigantic duplicate, substantially, of their joint image (upside down and every other direction reversed) on the retina of the observer's eye where he *sees* them.

Clearly, now, if from a point a long ways off, in the direction from the observer of a particular spot on the screen, a number of straight lines should be drawn that passed near to but on all sides of him, they would seem to come from the given spot on the screen, and each to run out from that spot in its appropriate direction, to the right, in the case of a line from the source that passed on that side of the observer, to the left, in the case of lines that passed on the opposite side, and similarly for all the others.

Hence these lines, however nearly parallel, would seem to spread out on all sides like the spokes of a wheel from its hub. That is perspective, and that is why the crepuscular rays, though parallel to each other, seem to radiate like the ribs of a fan from its pivot.

Evidently, too, if the straight parallel lines just run on and on, and one should look along them towards their destination they would seem to converge to a point diametrically opposite from their place of origin. And that is just what does happen in the case of crepuscular rays that start from the sun and pass by us, over head and on either side.

They may come from a setting sun, and, though parallel through their whole course, seem to consist of two great spreading spokes of light and shade, one, the "crepuscular rays," springing from a point in the direction of the sun, the other, the "anticrepuscular rays". Thus does the illusion of perspective often enhance the beauty and charm of Nature.

Fig. Anticrepuscular Rays.

Perspective also causes the nearly parallel beams of the aurora to appear to converge far more rapidly than they actually do, and, to an observer situated where he can look along them, seem to radiate in all directions from a common point of origin.

Another interesting perspective phenomenon is "Noah's Ark," a system of long, straight parallel bands of cloud that seem to start from a distant point in one direction, gradually separate to their widest spread overhead, and as gradually close again as they stretch away to a far-off point in the opposite direction—a colossal canoe (Noah's Ark) in the sky, narrow at both ends and broad in the middle. A straight stretch of railway affords the same illusion, narrow gauge a mile away in either direction, however broad-gauge where the observer stands.

Fig. Noah's Ark

Perspective also makes a whole sky-full of separate cumulus

clouds, though their bases of course are flat and all at the same level, look like the underside of a giant marble stairway, mounting from the distant horizon to the dome overhead. The base of each individual cloud simulates a tread in the great stairway, while the visible portion of its near side is the next rise.

A full display of this, the echelon cloud, or *cumuli en echelon,* is exceptionally beautiful, but quite unusual. Fragments of it, however, like portions of a ruined approach to some ancient temple that rose to the heavens, are common of summer afternoons, and always worthy of long observation by every contemplative lover of Nature.

Perspective, therefore, through the medium of clouds, creates some of the sky's best puzzles; but it also can queer us without clouds. It makes the full moon, for instance, look much smaller when well up in the heavens than it does near the horizon, and the query is: How? By causing the moon always to appear to be on the inner surface of that "inverted bowl we call the sky," a flat bowl, of course, whose rim, on the horizon, is much farther away than its bottom overhead.

Now, our daily experience tells us that of two spheres, for instance, that have the same actual angular spread at the eye, the more distant one is the larger of the two, in the same ratio as their distances from the observer. But the actual spread of the moon, as we know by instrumental measurements, is very nearly the same wherever it is in the sky, from horizon to zenith.

Hence, where it seems farthest away, that is, on, or close to, the horizon, it also seems largest, and conversely, wherever it seems nearest it appears smallest. This apparent change in the size of the moon, with position in the heavens, is a universal experience, an illusion from which no amount of astronomical knowledge can free us. Things will seem to be what they seem to be, however certain our knowledge that in reality they are something else.

As just explained, the full moon looks bigger to everyone when it is close to the horizon than when high in the heavens; and also to most people it looks just "so big" when well up, but they differ greatly among themselves as to the value of that "so big."

A very few say, about the size of a dime, which means that their perspective plane is only some seven feet away, the distance from one at which a dime seen flatwise just covers the full moon; and at that distance it will just cover it wherever it is, from horizon to zenith. Many say the moon, when high up, is about the size of an 8-inch plate. They instinctively think of the plate as being approximately 75 feet away.

Some say the size of an automobile wheel, which they clearly think of as being distant about 100 yards. These, of course, are just sample sizes for they range over every gradation, certainly from a dime to a wheel 20 feet or more in diameter. This is one of the easiest things in the world to get up an argument about, and the seemingly crankiest individual of all is the very occasional one who really sees things aright, and says therefore that the question: "How big does the moon seem to be?" has no definite meaning, since to him it has any size whatever, according to the distance of the comparison object.

He, indeed, is quite right, but few people are so mentally endowed that instinctively with them knowledge and logic have precedence over impression.

ATMOSPHERIC REFRACTION

We generally act on the assumption, or as we logically would if we did so assume, that light travels in a perfectly straight direction. That is why a marksman sights his gun at the thing he hopes to hit, not more or less away from it, as he would have to do (assuming the bullet to go straight) if light followed a devious course, or even a considerably curved one. For most of our everyday acts and occupations this assumption is entirely satisfactory, but not for all.

The Indian, for instance, knows that to hit a fish with an arrow he must shoot under it to this or that extent, as determined by its depth in the water and the slope of the arrow on reaching the surface. As a matter of fact, light travels in a strictly straight line only so long as the medium through which it goes is of the same substance and at constant density, or in the very special case where the light goes normally (not slopingly) out of one medium into another. In this special case the light keeps straight on, though at a different speed.

Now, although the composition of the air seldom differs enough from place to neighboring place noticeably to change the direction of a ray of light through it, density, incident to variations in temperature and pressure, often does differ sufficiently to be the cause of optical phenomena that range all the way from the normally unnoticed to the universally astounding.

One of these refraction effects to which commonly we pay no attention, and of which many people are even wholly unaware, is the complete rising of the sun and moon before they otherwise would even peep over the horizon; and their remaining up until after they would have been wholly down.

This comes from the gradual refraction or bending of light, that enters the atmosphere from the outside, into a more and more nearly perpendicular direction as it passes from rarer to denser air. Normally, for objects on the horizon this bending, then at its greatest value, is enough to lift an object more than the entire diameter of either the sun or the moon.

Evidently, therefore, the day begins earlier, and ends later, than it would if there were no atmosphere, to an extent that depends on the slope of the sun's course to the horizon, which in turn depends on the time of the year and the latitude of the place of observation. This lengthening is least, about 4 minutes and 40 seconds, where and when the path of the sun is perpendicular to the horizon, and longest at the poles where the day, in the sense of duration of possible sunshine, is increased by some 36 hours.

Another refraction phenomenon which few people seem to know anything about is the "green ray," or "green flash," as it also is called. This is the brilliant green, lasting commonly for but a moment, of the last tip of the sun as it sinks, or first as it rises, beyond a distant, bare horizon (a calm sea horizon is especially favorable) in a clear atmosphere. It is very beautiful and exceptionally easy to explain.

The extent to which light is bent out of its initial straight course on passing through the atmosphere varies with its colour; red is least bent, then yellow, green and blue, in the order named. Of these various colors, then, that reach the eye of the observer from a single speck of the sun, such as its last tip above the horizon, red, being the least bent, passes through the air below the others and by way of a lower point in the sky; blue, because the most bent, is the topmost ray; while green, having an intermediate value of refraction, or bending, lies between the two extremes, red and blue.

Violet, though even more refracted than blue, is not here considered because not enough of it gets through the atmosphere to be conspicuous.

Obviously, now, as the western horizon comes up (sun going down) incident to the rotation of the earth, it crosses the vertical fan of colored rays from bottom to top, and thereby cuts out from their total blend, which is white, first the red then, in rapid succession, the yellow, green and blue; the visual effect being an abrupt change from white to brilliant green, the green ray, that speedily fades away through blue when the air is exceedingly clear.

Usually, however, the air is so filled with dust, especially over land, that the last speck of the setting sun shows only red, the colour that penetrates a dusty atmosphere far better than does any other. Indeed very rarely is it so clear that it transmits enough of the blue light to show the exquisite phenomenon, where the sun disappears, of a gleaming emerald changing into a glowing sapphire.

Beautiful, exceedingly beautiful is the green ray. Then, too, there is the Highland legend to the effect that once seen it prevents all future deception in love. A closely kindred phenomenon to the green flash is the "scintillation" or "twinkling" and unsteadiness of the stars, mentioned by Aristotle more than twenty-two centuries ago, and ever since time and again puzzled over, until 1868-1869, when the astronomer, Respighi, finally cleared it up by many systematic observations and experiments.

Although the air varies in density approximately only with height, as a direct result of the consequent change in pressure, it actually is filled with innumerable small irregularities in this condition at all times and everywhere, owing to convections, incident to heating and cooling, variations in composition resulting from evaporation and condensation, and for still other reasons.

Now, these variations in the density of the air induce corresponding changes in both the direction and speed of a ray of light passing through it—give it such a devious and rapidly changing course that a fixed star, as seen through a large telescope, has a position nearly as definite and steady as that of a moth about a candle. In the twinkling of an eye its image is here, there and yonder, all over the field of view, so that the only way to measure its true position is to sight on the place where it isn't—the estimated mid-point of all places it seemed to occupy during the course of the recorded observation.

It is like locating a candle flame when all we can see is the moth flying about it. Obviously, therefore, when this jumping about of the light of a star is bad we have a case of "now we see it and now we don't," according as the light comes to the pupil of the eye or misses it. This is ordinary unsteadiness and colorless twinkling. A bright star close to the horizon, however, often shows a more interesting kind of twinkling, that in which it not only changes in brightness, but also rapidly passes through various colors without order or regularity—just fortuitously.

This sort of twinkling occurs, as stated, only in the case of stars near the horizon where the refraction is so great as to appreciably spread the light out into its several colors each of which follows its own course, so devious and so different from the others that first one and then another comes to the eye in both rapid and irregular succession; all due, of course, to rapid and irregular changes in the density of the air, incident to convection and other causes.

So much for celestial scintillation, but there also is *"terrestrial scintillation,"* as, for instance, of a distant are light, due to exactly the same conditions that cause the stars to twinkle.

The apparent unsteadiness and "shimmering" of objects seen

across a heated dry surface, also is a manifestation of atmospheric refraction, analogous to, and caused in the same way as, the scintillation of a distant are light, that is, by numerous and rapidly changing irregularities in the density of the intervening air, that refract the light this way and that, and thereby transmit it to the observer in such an unsteady direction that everything he sees by it appears to be all in a quiver.

Some people call this phenomenon by the very appropriate name *"laurence,"* in memory of St. Laurence who was martyred by roasting alive on a gridiron. It also is called *"optical haze,"* because although of a purely optical origin, as explained, it causes the outlines of objects to seem vague and indistinct, as they do in a hazy (dusty or slightly foggy) atmosphere.

As stated above, light coming in slantingly through the atmosphere, as from a star, gradually turns more and more nearly to the perpendicular as it gets into denser and denser air, or, in general, comes closer down to the earth. That is why the astronomer finds the rate of change in the direction of a star to be less than his nearly perfect clock says it should be.

The cause of this bending is the fact that the denser the air the slower the travel of light through it. Evidently, therefore, light that just skims the horizon in a level region, such as a calm ocean, tends to keep down to the surface, but it does not tend to do so strongly enough to succeed. It does curve down, but at a much less rate than does the surface of the earth. If this curvature were the same there would be no horizon, no place where earth and sky seem to come together; theoretically, at least, we could look straight forward clear around the world and see our own backs.

Our view, however, is not so extensive, as we well know, but the more a ray of light grazing, or tangent to, the surface is bent down, the greater the width of our view, or distance to our horizon, as seen from a position of given height. Therefore, the more rapidly the density of the air decreases with increase of height, or, what amounts to the same thing, the less rapidly its temperature decreases, or, better still, the more rapidly it increases, as it often does, with increase of height, the more pronounced is the down bending of a tangent, or near tangent ray and the farther one can see over the rim of the earth.

Hence, early in the morning after a still clear night, the time and circumstance of a strong temperature inversion incident to marked surface cooling, one's horizon is farther away, sometimes much farther, than it normally is, so that objects usually invisible (beyond the horizon) from the observer's position come clearly into view. They loom up, and the phenomenon therefore is called "looming." If instead of being afar off, the object looked at is rather nearby, a hill, say, half a mile away, and the temperature at the ground many degrees lower than it is at the height of 20 to 30 feet, as sometimes occurs, then the light from that object is strongly bent towards the vertical as it passes down through the cold surface air, so that to any one within that layer familiar objects look much taller than they actually are —the hill on which his home is located, and to which he then may be going, has mysteriously become a mountain, and its modest trees forest giants. Spooky, this "*towering*," as it is called, when seen for the first time, and still arresting no matter how thoroughly its cause is understood.

From the explanation of looming and towering just given, it is clear that if the density of the air should increase less rapidly than it normally does with approach to the surface, then a grazing, or tangent, ray will bend down less than it commonly does, and objects that usually we see sticking just above the horizon will be out of sight, beyond, or below it; and relatively near objects decreased in stature.

The first of these phenomena, disappearing beyond the horizon, is appropriately called "*sinking*"; the second, apparent decrease of height only, not of width, might, perhaps, be called "*stooping*."

This abnormal rate of change of the density of the surface layer of air, with increase of height, accounts also for two other striking optical phenomena, the "*superior mirage*," or appearance of an object as if seen in a horizontal mirror overhead; and the "*inferior mirage*," or seeming reflection of an object in a horizontal mirror under foot.

Thus, when the surface layer of air is much colder and denser than that just above it, in which with increase of height the temperature at first increases rapidly, an observer within this cold bottom layer sees an object some distance off in this same layer just as it ordinarily appears, and at the same time sees also just above it its duplicate turned upside down—a *superior mirage*—exactly as its image would appear in an overhead horizontal mirror.

Here light from the top of the object in question, that goes very slopingly into the warmer air above, is bent back again, at the same slope, into the bottom layer to the observer, while light from far down on this object that likewise reaches the observer by the way of the overhead warmer air enters it and leaves it more steeply; from which it follows that the parts of the object that are actually high appear relatively low in the sky, and the parts that are actually low appear relatively high, and the whole object therefore suspended in the air upside down directly above its duplicate on the ground.

When, however, the air near the surface is lighter than that just above it, as it often is over a sandy desert, especially of calm clear afternoons, a curious optical phenomenon, analogous to, and yet quite different from, that just described, frequently may be seen, the most common, most deceptive and most amazing of which is a "lake" where there is no water, reflecting islands in it, the shore beyond and the sky above.

To one a short ways up in the cooler air an object some distance away, and also above the heated surface layer, appears just as he is accustomed to seeing it, if he looks straight at it; but on looking below it he sees it clearly reflected, upside down of course, in a smooth body of water, where, as just stated, there is only desert sand. Strange and mysterious it seems, but the explanation is quite simple.

Light from that portion of the object which is just above the heated stratum may enter it very slantingly and then, owing to refraction that turns the ray towards the denser air, come up with the same slope that it had going down, and reach the observer as it normally would from an object exactly like the one it did come from, but directly beneath it; while light that comes to the observer from a higher point on the given object enters and leaves the warm surface-layer at a steeper angle, and therefore appears to come from a correspondingly lower point.

Hence under the given conditions one sees the actual object as and where he commonly sees it, and he also sees it as he normally would if it were at a lower level and turned upside down, precisely as it would appear in a horizontal mirror of great extent slightly below him.

One of the things seen by this refraction is the sky just as it appears by reflection from a body of water, hence the all but compelling conviction that we are looking at a real lake whose near shore is a short ways off, an illusion that is made still more convincing by the shimmering of the surface air, incident to strong heating, that produces the effect of rippling waves along the shore—a shore, however, that can never be reached, receding as fast as approached.

One of the most interesting things about this inferior mirage is the fact that nothing on the surface of the earth within it can be seen. That is why General Allenby ordered: "Cease firing," during a certain battle of World War I when a mirage completely hid the enemy from view.

The familiar road mirage and street mirage, in which we see people unconsciously splashing along or across a flooded highway (dry as dust), are only miniature examples of the inferior mirage—miniature, but intensely interesting and often even amusing; at least, the puzzled expression of one who never saw it before is amusing to the initiated.

REFRACTION BY WATER DROPS

One of the most familiar of all optical phenomena in the atmosphere is also the most beautiful, or, at least, one of the most beautiful—the rainbow, produced by the refraction of sunlight, or, rarely, moonlight, by water drops in the air.

When the sun is rather low at one's back, and shining brightly on a heavy shower in front of him, he sees in that shower two rainbows, an inner and an outer ("primary" and "secondary"), each a group of closely adjacent circular arcs

of the primary colors about a common centre, from which the straight line to the sun passes through the eye of the observer. The inner bow is much the brighter of the two, with red outer rim, paralleled by bands of yellow, green and blue in the order named.

Close inside the blue arc, and parallel to it are additional bows that show scarcely more than two colors, red and green, and that rapidly become closer together and fainter with increase of distance from the main, or primary bow.

Fig. Rainbow

These are called "supernumerary" bows—better, perhaps, supernumerary primary bows, being associated with, and in a sense a part of, the primary bow itself. Everyone that looks at a bright rainbow looks also at these supernumeraries, he couldn't help it if he tried for they are right there before his eyes, one to three or four, at least; and yet few people ever see them without first being shown, after which they always see them.

The colors of the secondary, or outer, rainbow, are in the reverse order of those of the primary, that is, red inside and blue outside with yellow next to the red and green next to the blue. There also are supernumerary secondary bows, and they too, like the colors, are reversed in position to those of the primary bow, as they are along its convex, or outer, side instead of the inner.

The sizes of these bows, or, rather, their lengths from end to end, vary greatly, both with the portions of the complete circles, of which they are parts, that are visible, and with their distance away, but their angular radii, or angular distance from the centre to any given one of the colors is always substantially the same, 42 degrees to the red of the primary bow, and 50 degrees to the red of the secondary.

In the brightness and purity of their colors rainbows differ greatly, one from another, owing to the size of the raindrops that produce them and to the uniformity, or lack of uniformity, in their size. Rather large drops of nearly uniform size give the brightest bows and purest colors.

When the drops are very small, fog drops, for instance, the bands are quite broad and their colors so indistinct that the whole bow is but little more than a faint white arc.

Another conspicuous rainbow phenomenon that comparatively few people seem to know anything about, or ever to see, is the distinctly greater brightness of the regions outside the bows than that between them. This, of course, is just an annoying puzzle until we know from any good book on optics how to explain it, after which it arouses in us a peculiar pleasure, the intellectual pleasure that comes of knowledge and understanding.

Sometimes we see a rainbow reflected in a calm surface of water. However, it is not the bow we see directly but the one that would be seen by an eye immediately beneath our own and as far below the level of the water surface as ours is above it. Sometimes, too, we see four rainbows in the sky at the same time, two primary bows, united at the surface of the earth, and most separated at their topmost points; and two secondaries, also united at the surface and most separated at their highest parts.

The inner of these primary bows and the inner of the secondaries are produced by direct sunlight, and the outer of each by light reflected from a calm body of water, and reaching the raindrop as though from a sun angularly as far below the water surface as the actual sun is above it.

Finally there are horizontal rainbows produced by a sheet of water drops, such as may be deposited by a drizzle on an oily lake surface, or by dewdrops on the grass of a well trimmed lawn. This latter phenomenon, the dew bow, is fascinatingly beautiful—full of gleaming rubies, emeralds and sapphires that transform, one to another, with every shift of the head, and strung with everchanging opals—exquisite, looked at by many, yet seen by few.

People alive to their surroundings—curious, like the small boy, to know what makes the wheels go round—ask many questions about the rainbow. Sometimes we are asked how far it is to a particular rainbow. To this question there is no answer in terms of a linear measure, such as feet or yards.

At best we can only say that its distance is just that of the raindrops that produce it, an indefinite value that covers the whole range from the nearest drops that contribute any portion of the colored light, to the farthest. Sometimes even the closest of these drops may be miles away, while in the case of a small fountain spray we can reach right into the bow and even grasp the "pot of gold" (yellow spot) at the rainbow's end.

Occasionally some one asks whether it is possible to go right under a rainbow's arch and come out on the other side. Often, too, whoever asks this question waits not for a reply, but immediately answers it himself with the emphatic assertion that he has done just that thing, and counters all incredulity we naturally might have by giving the exact hour and place at which he had this magical experience.

He may sincerely believe that he threaded the rainbow, but every physicist knows that such a thing is impossible—that the rainbow always is in front of one who has the sun at his back, and that it cannot have any other position. Of course, one might see a rainbow to the east, say, and then, if driving along an unfamiliar winding road, presently become so "turned around" that as he looks at the bow again he thinks it is the west, and therefore concludes that he has driven through it, just as he might under an arched bridge.

Now and then someone asks why rainbows are more frequently seen in the summer than during winter. The obvious answer is: Because the local shower in sunshine, necessary to the production of the rainbow, normally is caused by convection induced by a relatively high surface temperature, a condition that occurs far more frequently during the warm season, when the surface is strongly heated by sunshine, than in the winter.

"Why are noon rainbows so rare?" Is another legitimate question occasionally asked. That is easy. Since the centre of the rainbow, being exactly opposite from the sun, is just as far below the horizon on one side as the sun is above it on the other, therefore, even the uppermost tip of the primary bow would just graze the horizon when the altitude of the sun is 42°, the angular radius of this bow.

Hence, as the altitude of the sun at noon is greater than 42° everywhere except always in high latitudes and during winter in midlatitude regions, times and places where rainbows rarely occur at any hour, it follows that a noon rainbow is a rare sight at best, and even an impossible one (for the primary bow) anywhere within the Tropics.

Another popular question about the rainbow, perhaps the most frequently asked of all, is: "Can two people see the same bow?" The answer is "No." The rainbow is not a material object that, like a tree, for example, is fixed in position and visible from all sides. If we happen at any time to observe a rainbow so located that we can see the places on the ground from which it springs, or seems to spring, and then walk to one side we shall see what appears to be the same bow correspondingly shifts to new places of support, but we know that nothing material was so shifted.

Hence, the bow seen at the one place is different (made up of light from different drops) from that seen at the other place; and this holds for any two places however close together. Therefore, as the eyes of two people do not coincide they do not, and cannot, ever see the same rainbow.

REFRACTION AND REFLECTION BY ICE CRYSTALS

A ring around the sun or moon, way out from it, 22°, in fact, is a common occurrence in all parts of the world, though now and then some city dweller gets excited about it on "seeing" it for the first time, and phones frantically to know whether this terrible sign in the heavens portends war or the end of the world!

Strange enough, too, he, or she, seems disappointed when told that the worst it portends is falling weather, and not very certain of that. There are more than a dozen well known signs of this kind, halos, that now and then appear in the sky when covered with thin, high clouds of snow crystals. By far the most common of these is the "22° halo," a circle of 22° radius, around the sun or moon, red on the inside, followed by yellow and green, to blue on the outside.

It is produced by the refraction, by minute six-sided columns of ice, of light that enters one side-face and comes out at the second face beyond, after passing through the crystal in the shortest direction parallel to the face skipped. In this case the direction of the light is changed on passing through the snow crystal by about 22°, the red least and the blue most.

When there are myriads of such crystals in the air, pointing in every direction and turning over and over, the sky, whitish from scattered light, is illuminated also with this 22° circle of refracted light.

Sometimes there also are enough brightly reflecting vertical faces of snow crystals to produce a white circle, all the way around the heavens and everywhere at the same distance from the point directly overhead (the zenith) as the sun itself. Occasionally two colored patches, red nearest to the sun, blue farthest away, appear on this circle, or where it would be if developed, one to the right of the sun, the other to the left, and each just outside the 22° halo. These are the "22° parhelia" (side suns), or sun dogs.

That is why the white circle that passes through them, and through the sun is called the "parhelic circle." The flat, horizontal, or nearly horizontal, bases of prevailingly upright columnar snow crystals also reflect the sunlight, or moonlight, that may shine upon them, and thereby produce that white, vertical column which, in conjunction with the parhelic circle, forms the well known "cross" whose centre, or intersection, is at the sun, or moon.

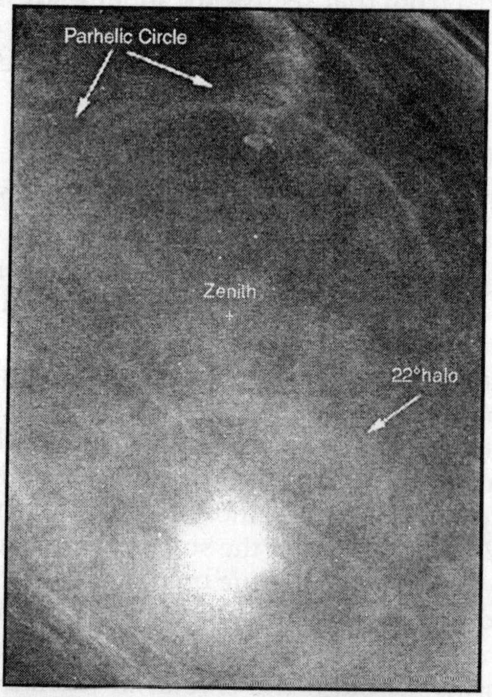

Fig. Halo of 22°, and Parhelic Circle

When myriads of tiny ice columns, turning in every direction, have flat ends normal, or perpendicular, to the sides the "46° halo" (halo of 46° radius) is formed around the sun or moon by light that enters a side, or base, and comes out through a base, or side, in such direction that it is equally inclined to each, and of shortest path through the crystal from the point of entrance, consistent with this condition of equal incident and exit slopes.

Occasionally the bases of these crystals are prevailingly horizontal, in which case, when the sun is about 22° above the horizon, the "circumzenithal arc" may be seen, a brightly colored ribbon that covers, roughly, one quarter of a complete circle about the point directly overhead, the zenith, as its centre, convex and red towards the sun, blue to violet on the inside.

This is much the most brilliant of all the halos, so brilliant, indeed, that it often is mistaken for a bright rainbow. It is startlingly beautiful, but to see it one must be on the *qui vive*, for its duration at anything like its best is only a few minutes.

As already stated there are a number of halos in addition to those just described, but they are rarely seen, and therefore no attempt will be made to explain them here—the game does not justify the candle.

DIFFRACTION BY CLOUD DROPLETS

We know from experiments and observations that water drops, of whatever size, reflect from their outer surface a portion of any light that may fall upon them; that some of the light that penetrates into a drop is immediately refracted out again through its farther side; and that the rest gets out only after one or more internal reflections.

This accounts for the rainbows, for instance, and several other phenomena. But this is not all that water drops do to passing light; they affect some of it that doesn't even touch them—drag it in behind them and illuminate their shadows with it.

This curious effect is called "diffraction," and when it is produced on sunlight, or moonlight, by a great number of droplets, all of the same, or very nearly the same, size, the result is a series of colored rings, called a "corona" red on the outside, blue on the inside, around and rather close to the luminary, one or two to several degrees, as determined by the size of the droplets, in the sense that the smaller the droplets the larger the rings. When the droplets are quite small there may be two, or even three, sets of these rings around the sun or moon. Such rings around the moon can, of course, be observed directly, but to safely and comfortably see those about the sun one should always use strongly colored glasses, or else look at their reflection in something that returns only a small fraction of the total light, such as very dark glass, ink colored water or even a muddy puddle.

When the droplets are exceedingly small, as they must become in slowly evaporating thin clouds, the corona appears as only splotches of iridescent cloud, very beautiful but apparently rarely seen. When the cloud droplets are very irregular in size they produce only an "aureole" or brownish glare close to the sun or moon.

COLORS OF THE SKY

The commonest of all atmospheric optical phenomena, so common, indeed, and so universal that we seldom give it any attention, or even notice it, is one of the most beautiful, and the one that went longest without explanation—the blue of the sky. The other sky colors also are beautiful but they are neither so extensive nor so persistent as is the blue, being mainly restricted to the hours and general regions of sunrise and sunset.

In earlier times, and indeed until but a few decades ago, all efforts to explain the colour of the sky were confined to attempts to account for the blue, and all were futile. The first hint at the correct explanation was gotten from Brücke's experiment in 1853, which showed that a transparent medium rendered turbid by very fine particles appears blue when illuminated with white light. In 1869 Tyndall made numerous experiments of this kind and showed that not only is the scattered light thus obtained blue, but also in other respects exactly like the light of the sky.

Hence, as the blue light thus artificially obtained is the result of selective scattering out of the white light (scattering a larger portion of the blue than of any other colour) by minute particles, therefore, the sky, we infer, gets its colour in the same way, by selective scattering of the blue out of white sunlight.

Then in 1871, more completely in 1881, and still more fully in 1899, Lord Rayleigh developed the scattering theory of the blue of the sky and expressed it so concisely in terms of mathematical equations that this ages-long puzzle is now considered to be definitely solved. From theory and from experiment it appears that of all the colors in sunlight, red, orange, yellow, green, blue, indigo and violet, the first is the least scattered, or least depleted by scattering on passing through a given mass of the atmosphere, and the others more and more scattered, in the order given, on to the violet which is so rapidly depleted that but little of it reaches sea level.

Clearly then the dominant colour of the clear sky, in any particular direction, depends on the amounts of the lights of the several colors that pass into the air that lies in that direction; the fraction of each that is scattered by this air; and the portion of that fraction in turn that gets down to the surface of the earth.

Thus, though a larger portion of the violet light is scattered than of any other colour, it is relatively so weak initially, and so rapidly depleted on its course to the observer that only from very great heights does the sky look distinctly violet. Again, though sunlight is rich in red, it is comparatively too little scattered ordinarily to make that the colour of a clear sky. In this way, that is, by considering the initial intensity of the light of each colour, and the extent to which it is scattered by the atmosphere, we soon see that normally the dominant colour of the sky must be blue.

We see also, from these considerations, that sunlight that has gone a long ways through the atmosphere, from west to east, for instance, at some distance above the surface, must lose most of its blue, indigo and violet before getting across, and therefore give a clear twilight sky over a considerable area well above the western horizon a distinctly greenish cast, paralleled by yellow to orange at the next lower levels. Similarly, the light that nearly grazes the surface is so relatively rich in red by the time it gets halfway across that the sky takes on a ruddy glow for some distance above the place where the sun just went down, or, similarly, where it soon will rise.

And since this grazing sunshine is dominantly red, therefore broken twilight clouds and the scattered clouds of dawn, illuminated by this light, are ruby red to rosy hued—glowing glories of the sky that, savant to savage, holds all the world in awe and admiration.

Chapter 4

Climate, Present and Past

It commonly is said that the climate of a given place is its average weather, yearly, seasonal, monthly, or by whatnot other given period; a definition easily remembered, but wholly useless because weather, being a composite of temperature, humidity, precipitation, sunshine, and all the other meteorological elements, is not averageable. It is absurd to talk about averaging even two of these elements, such as temperature and humidity, for instance.

They may exist together, but they cannot be added, the one to the other, into some sort of a whole that can be numerically divided, and so there can be no such thing as an average hot-wet, much less an average hot-wet-sunshine-pressure-wind-etc. The expression "average weather," then, does not make sense, at least not the kind that leads to numerical evaluations, and therefore classification, of the thing itself, weather.

If, however, we should agree that "average weather" shall mean average temperature (over a specified period), average humidity, average pressure, etc., we then would be no better off, for there could be no satisfactory measure of the departure of the weather from this average as a result of a given departure of any one of the elements from its average.

Another definition of the climate of a place is: "The average run of its weather." This definition is better than the first in that it specifies that the average is to be a continuous one, however variable, just as the ever-changing weather goes on and on without beginning and without end.

But for statistical purposes this definition, too, is out of the question, for it implies doing the impossible—adding together the weathers of many consecutive New Year's days and finding their mean value, and similarly for all the other days of the year. Of course a shorter or a longer unit period than the day might be used, but whatever the period the same two impossible operations are implied, namely, adding weather to weather, and finding the mean value of this sum. The best that can be done to determine the average run of the weather of a given place is to find from observations covering many consecutive years its average temperature, average humidity, and average every other element, for each day, hour, or other definite period, throughout the year.

This is valuable information but it does not contain all the pertinent facts. It leaves out critical values, extreme values and the rates of change of the several elements, conditions that often are of paramount importance. Clearly, therefore, the only complete account of the climate of any place is the full history of its past weather through the particular period under consideration.

Partial histories of this kind (partial, since only some of the elements are recorded) are now available for a great many places, in the making at many others, and contemplated for innumerable still others, but the very vastness of these rapidly accumulating data, already tombed in tons of tomes, precludes the publication of all that is available about the climate, or rather the multi-various climates (each station has its own climate) of any considerable region.

So much, then, in general as to what the climate of any particular place is, but we also want to know how it got that climate—what the factors are that determine what the climate of this place shall be. Not only is this information intellectually satisfying, but also it is practically useful, because from it the climates of many places for which there are no records can be estimated with a very helpful approach to accuracy.

Obviously the most important factor of all that jointly make the weather, and hence the climate, of any place is the rate of output of radiant energy by the sun, because that determines the temperatures of land, water and air; fixes, in general, the amount of evaporation and its counterpart, condensation; controls the winds and regulates in greater or less degree every other weather element.

Any considerable decrease in the intensity of the solar energy just outside the atmosphere, a decrease of ten per cent, for example, if long continued, would have the disastrous effect of spreading a mile-thick sheet of ice over a large portion of the earth's surface; while a similar increase in that energy, also if long continued, would be equally disastrous through the rapid melting off of all remaining ice sheets, the consequent considerable rise of the ocean (150 feet, by a conservative estimate) and drowning out of low lands and most coastal cities, and the spread of arid conditions to half, or more, of all the now fruitful areas in the mid-latitudes.

The rate at which the sun gives out this energy may have varied considerably from time to time during the geologic ages, but there is no proof that such changes did occur, in fact such evidence as we do have is quite to the contrary, to the effect, namely, that during all the hundreds of millions of years of geologic time the output of solar energy was at least roughly constant.

We know that ever since accurate measurements have been made of the solar constant (intensity of the solar radiation at mean solar distance), that is, ever since about 1910, these changes, though apparently of frequent occurrence, certainly have been small and of short duration. Therefore it may be assumed, as a working hypothesis, that the rate of output of solar energy is, and for ages has been, approximately constant.

Of course, though, it is the rate at which this solar energy is absorbed by the earth, how absorbed, and how distributed, that really count in the making of weather and the fixing of climate. Therefore if the distance from the sun to the earth should change, so also, if the sun were constant, would the intensity of the solar energy at the upper limit of our atmosphere.

But although this distance does vary by about three per cent through the year, from least, early in January, to greatest, early in July, its mean value is so very nearly the same year after year that no appreciable change in climate can be attributed to a change in the solar distance. It is true that the annual change in the distance of the earth to the sun varies slightly through long ages, over, roughly, a 100,000-year cycle, but that period does not fit any known climatic change, and therefore certainly has not been the cause of these changes.

But our distance from the sun is not the only thing pertaining to the earth's orbit that not only conceivably could, but actually does, vary. Two of these, the obliquity of the ecliptic, and the excentricity of the orbit, as the astronomer calls them, are of climatic importance. At present the obliquity, or inclination of the earth's equator to the plane of its orbit, is very nearly 23½ degrees, the amount the sun every year swings north and south of the equator.

Clearly, if this swing were several degrees greater, and the other astronomical factors unchanged, the summers of middle and high latitude regions would be warmer than they now are and the winters colder; and if the swing were several degrees less, winters and summers generally would be less contrasty than they are at present. How much this swing of the ecliptic can change—has changed and will change—is not known, but it is known that it can not exceed more than one or two degrees in a million years, and therefore that it has not been an appreciable contributing cause of any climatic change that may have occurred within historic time.

The other variable in the earth's orbit, its excentricity, or, essentially, the departure of its elliptic path from a true circle, varies too slowly to be considered in connection with any climatic variations that may have occurred within the past 10,000 years, at least.

Obviously, when this flattening of the orbit is considerable it must affect the climate of every place, and astronomers say that the maximum, or aphelion, distance of the earth from the sun can be as much as one and one seventh the perihelion, or minimum, distance. If, now, while the earth's orbit is most elliptical the earth itself should come to the position of greatest distance from the sun, aphelion, during mid-winter of the northern, or land, hemisphere, it is quite certain that the mid and higher latitudes of this half of the world would then accumulate more than its wonted amount of snow and ice.

This occasional large excentricity of the earth's orbit, then, might be, and it has been assumed to have been, the cause of the ice ages, the long periods when much of the surface of the earth was covered with thick sheets of ice. However, this pleasing theory, essentially the theory published by the astronomer Croll in 1864, does not fit the facts.

Since the date of aphelion (greatest distance from the sun) runs through all the year, January 1 to December 31, inclusive, in the course of about 21,000 years, we would expect, according to this theory, that there would be extensive glaciation and deglaciation in the course of every 21,000 years.

Geologists, say, however, that the ice ages had no such frequency, but were of much greater duration and separated (there have been several such ages) by vastly longer periods. This theory also requires that the northern hemisphere and the southern must have glaciation alternately, whereas all the available evidence indicates that their glaciations were simultaneous.

Finally the earth's orbit has been nearly circular since around 50,000 years before the last great ice sheets had reached their maximum extents. Apparently, therefore, variations in the maximum and minimum distances of the earth from the sun have not been the controlling causes of either the great climatic changes of the geologic past, or of the relatively small changes that have occurred during the last 10,000 years.

There is, however, one other possible astronomical or celestial cause of climatic changes on the earth—nebulae along the course of the solar system through stellar space. But this pleasing hypothesis is untenable since it would account only for cooler periods than the present whereas the geologic evidence shows that the periods when the earth was decidedly warmer than it now is were far longer than those when it was cooler. If, then, the heavens fail us, as they seem to do, in furnishing an adequate cause for marked and long continued changes of climate, we obviously must consider how the earth, given a constant annual budget of solar radiation, can regulate, and presumably has regulated, her own climates.

One of the most obvious of these ways, a way that has been much written about, would be to shift her own skin around and thereby, if the movement were north or south, change the *klima* (Greek), or poleward slope, of the shifted surface. Thus, if all Canada and half the United States, for instance, had centered around the North Pole, they might have gotten a coating of ice something like that which still covered them 30,000 years ago.

And if Greenland had come down to latitudes 30° to 40° she might have acquired a flora like unto that which her fossils show that once she did have. This theory looks beautifully simple, but when one tries to so work it out in detail as to fit the known facts he soon is confronted with difficulties greater than those he is trying to avoid. Therefore, for the cause or causes of climatic change we perforce turn to those factors of climatic control that we know themselves have changed and, in some cases, why they changed.

Among these is the extent and composition of the atmosphere in respect, especially, to the two gases that are most effective in absorbing radiation, water vapour and carbon dioxide. Of these, the latter interferes but little with the incoming solar radiation but is moderately absorptive of the outgoing earth radiation.

Clearly, therefore, the greater the amount of this gas in the atmosphere the warmer, in general the surface of the earth, and, conversely, the less there is of it the cooler the surface. However, these effects are not nearly so great as at first one might suppose they would be, because the water vapour always present in the air absorbs rather effectively the same radiation as does carbon dioxide, so that variations in the quantity of the latter are compensated for in the main by the former, in the sense that the less the radiation taken out by the first the more the second has a chance to absorb, and the more, in general, it does absorb.

As is well known the amount of carbon dioxide in the air is in a continuous flux, constantly being increased by combustion and by the outpouring of volcanic gases, and as continuously decreased by plant growth and by the formation of bones and shells. Furthermore, the ocean is a most efficient regulator of the amount of carbon dioxide in the air—the colder the ocean the more carbon dioxide it absorbs, and the warmer it is the more it gives out.

A world wide decrease of tempera- ture then should lead to a still greater cooling by causing the ocean to remove from over the earth a portion of one of its protecting blankets; while the warmer the earth the still warmer, to some extent, it must get owing to the consequent addition by the ocean to the material of this protecting cover.

But, as already stated, it seems unlikely that variations in the amount of carbon dioxide in the atmosphere ever effected appreciable changes in the climates of the earth; and therefore that it itself either is, or ever has been, an important factor in the control of climate.

But of that other absorptive gas, water vapour, the story is very different, as it has a far greater control over the temperature of the air and the surface of the earth, and therefore on weather and climate, than have all the other gases of the air combined—all the remaining 99.73% of the mass of the atmosphere.

Like the witches of ancestral days it can wholly change its form, change from a gas to a liquid or a solid-cloud of droplets or snow crystals—and thereby become more effective than ever. So long as the water in the air is all vapour, or gas, it does not greatly decrease the intensity of the incoming solar radiation, but as soon as it condenses into a cloud, whether of droplets or crystals, it wholly repels and sends back to space, without its having accomplished anything whatever.

A large portion of this radiation (78 per cent when the cloud is fairly dense) that otherwise would have reached the surface of the earth and raised its temperature, and that of the adjacent air, to a somewhat higher degree. This loss of solar radiation is, of course, confined to the daytime. There is, however, another action of both vapour and cloud that operates continuously, both day and night, namely, the conservation of much of the heat of the surface and of the lower air.

The water vapour absorbs a large portion, perhaps of the order of one half, of the outgoing radiation, and even a larger portion where there are no clouds to take their big share. This absorbed energy appears as heat, which in turn is nearly all reradiated, half of it back towards the earth, and half on to higher levels, where in great measure it is again absorbed and reradiated, and so on, step by step, until a level is reached beyond which there is but little water vapour left, with, at every equal increase in height a decrease in the proportion of the radiation absorbed and an increase in that transmitted.

In any case, the water vapour alone sends back to the surface of the earth, where it is absorbed, a large portion of the energy which it itself had first emitted. This, of course, keeps the temperature of the earth higher than it would be if there were no such return of radiation.

The gathering of clouds only increases this absorption and

reemission of earth radiation, for in respect to it they are nearly "black" bodies, absorbing virtually all this sort of radiation that comes to them, and giving out radiation of this same sort almost as abundantly as is theoretically possible for any object to do at the same temperature.

Hence low clouds, being but a little cooler than the earth beneath, send down to it only a little less radiation than it itself gives off; and as well nigh all this return radiation is absorbed the temperature of the earth under any low cloud falls but slightly during even a long night.

When the clouds are very high, however, they normally are too thin to be first rate absorbers, or radiators, and also so cold that even if dense their radiation would be only around one third as intense as that of the earth beneath.

Very high clouds, therefore, could not greatly decrease night cooling, even if the air beneath them were wholly moisture-free, and all the less can they decrease it when the radiational interchange between them and the surface has to be, as it does, mainly step by step through the highly absorptive water vapour.

Clearly, then, the amount of water in the air, whether as a vapour, as liquid droplets or as solid crystals, is a factor of the greatest climatic importance. Naturally, therefore, we ask whether the amount of water in the air varies, and, if it does, how it varies and to what extent.

It does vary, obviously: Condensation takes water out of the atmosphere, irregularly both as to time and place, while evaporation continuously adds water to it. *Tersely:* Foul weather dries the air; fair weather dampens it.

Terser: Rain dries the air; sunshine wets it. However there is a definite limit to the amount (mass) of water vapour that can exist in a given volume at a given temperature, a quantity that rapidly increases with increase of temperature up to the boiling point. In general, therefore, the warmer the water vapour, or the air through which it is diffused, as they both must have the same temperature, the greater the amount of water in the air, for there is an abundance of water that may be evaporated—oceans of it.

Presumably, therefore, there was more water in the air during the warm geologic periods than there was through the ice ages. However, it does not follow that when this atmospheric water load was greatest, precipitation also was greatest. In fact it all but certainly was the other way around, since the ways for getting water out of the air, also essential to precipitation, doubtless were more active when the world was cool than when it was warm, as explained a little farther on.

Evidently, too, the large water content of the atmosphere during the warm periods aided materially in keeping the temperature of the earth relatively high, and its change with latitude comparatively small; while the thin vapour blanket over the earth during the cold periods allowed it to get still colder, especially when the nights were long (during the winters), and the difference between the temperatures of high and low latitudes to become more pronounced.

Another atmospheric content of at least some climatic effect is dust, of whatever origin, though that which is blown into the very high atmosphere by violent volcanic explosions seems to be of most importance; largely because it can remain suspended far above the highest clouds for many months.

When this dust, often consisting of flaky material floating flatwise, is so coarse that it depletes both sun and earth radiation by reflection and absorption chiefly, rather than by scattering, it produces a greenhouse effect and increases the temperature of the earth beneath, since such material is more absorptive of the latter (earth radiation) than it is of the former.

But when it has the normal size of the volcanic dust that floats for months, or even years, above the outermost clouds, that is, when it is only three quarters of one ten-thousandth of an inch across, it is of just such degree of fineness that it must deplete passing solar radiation to a much greater extent than radiation from the earth, and thereby produce an inverse greenhouse effect, by virtue of which the air and the earth become colder than they otherwise would be; an effect that was conspicuous in 1783-86.

Following the terrible explosion of Mount Asama in Japan; 1815-16, when we had the "year without a summer" here in America, after the violent explosion of Tomboro; 1883-86, after the destructive explosion of Krakatoa, that painted the sky in gaudy colors for two or three years; 1912-13, following the explosion of Katmai in Alaska, and doubtless after thousands upon thousands of other such explosions, back through all the earth-recorded ages.

How great the effects were of any of these prehistoric explosions we do not know, but we do know that there were times of great volcanic activity, and we know also that one or two score explosions, end on, of the Asama and Krakatoa type at intervals of several months to a year or so, would disastrously chill the climates of mid latitudes.

At any rate, volcanic dust now and again in even recent times has been an appreciable factor in climatic control, and presumably therefore always has been such a factor throughout geologic time-fitful and irregular, but occasionally so frequent as to keep the higher atmosphere dusty for many years together, and thereby to chill the whole earth beneath, not greatly, but enough perhaps (a very few degrees would suffice) to be disastrous.

As already stated, it is not certain that much of this sort of thing ever did happen, but it is certain that some of it did, and certain, too, that all of it is within the limit for volcanos to effect; and what they now could do, through the past ages they probably occasionally did do.

But there are still other things that greatly affect the weather and thereby help to control the climate. One of the more obvious and effective of these is the elevation of the land surfaces. The higher these surfaces, and their heights have greatly changed during the geologic ages, the colder, in general, the air at the same latitude, and also, up to a considerable elevation, the greater the amount of precipitation, at least on the windward side.

Thus, a high mountain range across the path of any wind necessarily causes that air to ascend, whereupon it cools by expansion, often to temperatures low enough to induce precipitation, mainly on the windward slopes, after which it flows down the opposite side of the range, becoming warmer with descent and drier too, not in respect to the actual mass, or weight, of water vapour present per cubic foot, say, but in respect to the amount necessary to saturate that cubic foot.

Hence wherever such a mountain range lies across the normal course of humid air, the climate on its windward side is certain to be rainy, while that to the leeward, even to long distances, tends to be arid. Furthermore, where the windward slope is steep, precipitation is apt to be superabundant—concentrated onto a narrow, drowned belt—whereas a long, gentle slope would get, from the same winds, moderate and altogether productive rains, and a radically different regimen of weather and resulting climate, as exemplified at many places over the earth.

The distribution of land and water, a condition that has undergone many great changes, is another controlling factor of climate. Just what the climates would be under this or that specific distribution of land and water no one can say in all details, but it is obvious that if, for example, the land and water were arranged in parallel east-west belts around the earth the north-south circulation of the oceans would be greatly reduced; high latitude regions would be colder than they now are and the equatorial regions warmer.

Evidently, too, if all the great deserts of the world were exchanged with equal oceanic areas marked climatic changes would result. And so, without going into details, one can see, as self evident, that certain changes in the distribution of land and water would greatly alter our climates, and therefore must hold that any considerable variation of the former must entail greater or less simultaneous changes in the latter.

The circulation of the atmosphere, or directions and strengths of the winds, is another obvious control factor of climate. Experience tells us that winds directly from cold regions have lower temperatures than do those from warm regions. We know from observation, too, that winds from over a warm ocean are more humid than those from a desert and provide more rain.

It also is known that up-slope winds generally furnish more cloud and more rain or snow than those that flow down-hill, and we understand why. Obviously, too, on-shore winds bring with them different weather from that carried by off-shore winds, and leave a different climate.

Clearly, also, winds carry the water evaporated at one place, together with its heat of vaporization, to some other, often far away, where the rain comes down and the heat goes free. Then, too, there are certain regions over which the northsouth movement of the air, the movement that by and large is an equalizer of world temperatures, is far greater than it is over others of even the same latitude, and that gives to these places quite different regimens of weather, and of course different climates.

Up the Mississippi Valley, for instance, is one of the routes most followed by tropical air, and down it a favourite course for polar air. Hence all that region experiences frequent general storms, and normally receives an adequate amount of precipitation fairly evenly distributed through the seasons; whereas, over western Texas, and on to the Pacific coast, such interzonal winds are relatively infrequent and feeble, and therefore the climate of all that area semiarid, and otherwise very different from that of the great valley over which the warm winds and the cold fiercely conflict in their struggle for room to pass.

Furthermore, since the winds of today are largely affected by the heights, trends and extents of mountain ranges, and by the areas and elevations of continents, it is evident, leaving out all other factors, that with the rising and the sinking of the mountains through the ages, and the spreading and contracting of the continents, the wind systems and the climates they determine also underwent profound changes.

Yet another great and effective distributor of heat and cold, and therefore an important factor in the control of climate, is the circulation of the oceans, by which the warmer water travels over the surface to higher latitudes where the air is made warmer than it otherwise would be, laden with more moisture, and the climate made comfortable; while the colder flows to lower latitudes, partly along eastern, high latitude shores, but mainly far beneath the surface where its movements are slow.

This overflow of the warmer water and underflow of the colder is owing, in part, at least, to the former's being lighter, by virtue of thermal expansion, than the latter.

To be sure the density of sea water is increased also by evaporation, which correspondingly increases its salinity. This is very effective at latitudes 25° to 35°, or thereabouts, over the oceans, where, owing to relatively clear skies, evaporation is vigorous, and where, due to the sinking of the warm surface water, incident to its increased salinity, the temperature from a short ways down on to great depths is higher, level for level, than it is either towards the equator or in the direction of the adjacent pole.

Nevertheless, the density of sea water, and hence, in part, its circulation, is altered mainly by changes in its temperature rather than by variations of its salinity. Presumably, though, neither the differences in temperature nor the differences in salinity, nor both combined, are the major causes of the ocean currents, but rather the combined drags of all the surface winds.

There still is one more important factor that contributes to the control of weather and climate, namely, the nature of the surface covering. When this is either water, or growing, leafy vegetation it contributes abundant moisture to the atmosphere and thereby keeps down the local temperature at the place of evaporation, and raises it at distant places where the vapour is again condensed.

On the other hand, where the surface covering is bare sand, or other dry material, no heat is lost by evaporation and the temperature of the surface layer, and of the adjacent air, becomes much higher in the sunshine than it otherwise would, and lower during still clear nights.

Again a snow surface, a surface that sometimes covers, one at a time or even all together, roughly, half of North America, half of Asia and most of Europe, produces a very different effect on the weather and climate than does either an all water surface or a dry sand surface. It is a nearly full radiator, or "black body," for the portion of the spectrum appropriate to its temperature, and therefore loses heat rapidly under a clear night sky, when, in the absence of clouds the return radiation is much less than the outgoing.

This entails also a considerable drop in the surface temperature except when the turbulence in the air, incident to a moderate wind, is sufficient to force a considerable depth of the lower air, instead of just a thin layer, to share in the loss of heat.

At night, then, the temperature of a snow-covered region commonly is below the freezing temperature. Nor by day can its temperature be any higher than that of melting ice because ice cannot be heated any hotter—more heat melts some of it but leaves its temperature unchanged.

Where the snow is deep, porous, and has a lower temperature than the freezing point, conditions that obtain at many places in the early spring of the year, there the water from the melted top snow trickles down through the colder snow beneath, itself gradually freezing as it goes and at the same time giving out enough heat (heat of fusion) to raise the temperature of more than 300 times its own weight of the colder snow by one degree.

Clearly, then, the effect of the absorption by the surface of a deep layer of snow of any considerable amount of heat is to melt some of the top snow, and then with this water, as it trickles down and refreezes, to bring all the snow to the same temperature—the temperature of melting ice.

After this, any further absorption of heat by the surface sends water to the ground beneath where it is either absorbed by the earth or runs off over it to appear at some lower level, as, or like unto, a glacier stream.

At no time, then, day or night does, or can, the surface of a snow field have a higher temperature than the freezing point, though it may, and on calm clear nights commonly does, become appreciably, to much, colder. Hence within such a snow covered region, and also to some distance to its leeward, the air always is cool to cold, and the climate a cold one. In short, the covering of any region, snow, water, vegetation or bare sand greatly affects its weather and goes far to determine its climate.

Although the climate of any place can be known without recourse to a well-kept series of meteorological observations at that place over a continuous period of many years, nevertheless from a knowledge of what factors control its weather, and their relative values, the type of its resultant climate can be inferred close enough for many purposes, as may be shown by the following, or any other, examples from widely scattered places.

To start with one of the most pleasant of these, the island of Oahu: Here the temperature of the water for hundreds of miles in every direction is mild. Mild, therefore, should be, and is, the temperature of the air at Honolulu, Waikiki, and whatsoever other place over the island, with never a frost and never a freeze.

The prevailing winds, being trades, are of moderate strength and nearly constant direction, and having traveled a long ways over water of moderate temperature should, and do, carry a large amount of water vapour, and therefore should, and do, afford superabundant precipitation on the tops and windward slopes of the mountainous interior, but comparatively little on the leeward slopes, and even less on beyond.

And so it is, gauges up on the higher peaks back of Honolulu have caught, on the average, 140 inches of rain per year, and others, a few miles away, on the mountains to the northwest 240, while the yearly average at the Weather Bureau's station in the city, to the leeward of the Koolau range, and therefore in its rain shadow, is only about 28 inches, and around Pearl Harbor, a few miles farther on from the air drying mountains, but half of that in turn.

That is, the windward side of the island and the higher portions of its mountains on all sides have a greatly superabundant rainfall, while the cane fields on the leeward side require considerable irrigation, since the rain they get is not only insufficient in amount, but, being incident, chiefly, to the occasional kona, or southwest wind, induced by a passing cyclonic storm, also too irregular in occurrence to be highly effective in the promotion of plant growth.

In short, from its location, and the factors that make its weather, the inference is immediate, and correct, that as to temperature its climate everywhere is mild, while as to precipitation it ranges the gamut from drowned to desiccated, or at least from five times as much as there is any need for to not half enough.

Nor is Oahu alone in having this made-to-order climate, a climate that affords plenty of rain back on the vegetable garden and a flood of sunshine on the flowers in the front yard. It is so with every similarly situated island, such as Puerto Rico which is in the path of the north-east trades of the Atlantic, and therefore has more than an abundance of rain on its windward, or northern side, and at the same time decidedly less than is needed on its leeward or southern side.

Here, too, the location of the island, well out in a warm ocean, and its moderate elevation must, one would infer, and they do, insure a moderate temperature at all times for its every portion with complete freedom from frost.

The coast of Norway also furnishes abundant material for the comparison of deductions from known factors with the data of observation. Here, incident to the latitude and the rotation of the earth, the prevailing wind is from the west, over an extensive body of water tempered in its upper portion by the Gulf Stream drift. At Bergen, typical of the coast, with steep mountains rising almost from its backyard, the air should never be, and never is, oppressively warm, nor unbearably cold.

In fact, despite its high latitude, 60° N., the same as that of Leningrad, its average January temperature is above the freezing point—a great contrast with that of many other places of about the same polar distance, such as Okhotsk, 60 miles farther south, where it is 11 degrees below zero.

And from the mountain factor, the presence of a rather high range near the coast, up which the well-nigh saturated ocean air necessarily rises as persistently as the west wind blows, it is evident that all this coast must be forever drenched in drizzle —when it isn't snow. At Bergen, again for example, the town that father and son Bjerknes have made to bulk large on the meteorological map, it is said to rain every day, most—that is, two days out of three, on the average.

In fact they say it is dangerous there to take a horse out on a busy street any time when it is not raining, as he almost certainly would get scared on seeing people going about without umbrellas. Up on the coastal mountains, and we don't have to climb far to find it, there is snow aplenty, while in the valleys beyond the precipitation is much less than it is on the windward, sea-facing slopes, and the temperature much lower.

Obviously enough, then, the climate of this coast must be cool to cold and rainy, inimical to the vine and the tree. A harsh climate it is, but not too harsh for the production of a fine and sturdy race. This will do for many purposes, but if details are needed one must have records from the place, often even the exact location at that place.

If, again to be definite, that place happens to be Bergen, then the location within the town must be specified, and the same thing is true, to a greater or less extent, of every city, town, village and country side on the face of the earth. However, a knowledge of these local variations must be left to the local people; they are too numerous to put on record, and besides most of them only interest the native any way.

Another of the many places that have easily predictable climates, is the great Amazon Basin. This region is wide open on the Atlantic side to the prevailing easterly winds which, because they have come a long ways over a tropical ocean are both warm and heavily laden with moisture.

Furthermore, since rank tropical vegetation covers nearly all this abundantly watered warm region, it follows that nearly as fast as watercomes down as rain it rises again as vapour to rain farther on, and again farther on, to the very foot hills of the Andes, where because the air is there forced to rise, the rain is heavier than ever.

There is no surprise, therefore, in the fact that the annual rainfall over some 2 million square miles of this basin would, if it were all cought of none of it got away, cover the whole of that enormous area with a lake six feet deep. Nor is there any surprise, either, in the fact that, as an Irishman might put it, the Amazon is the wettest river in the world.

Obviously, since the wind keeps blowing up the Amazon Basin it must somehow get out again, which it does chiefly by dividing into two parts, one of which goes over the Andes, losing so much of its moisture on the way that it comes down on the Pacific side "dry as a bone."

The other portion turns south, and after climbing and moderately watering the uplands of southern Brazil, merges with the prevailing westerly winds. For India also it is easy to determine roughly, and yet close enough for general purposes, what the climate must be. Here is an extensive low-lying area, open to a warm ocean to the south and southeast, but bordered by high mountains in the other directions.

These high mountains, the Himalayas, especially, to the north, protect this region from the cold winds and storms of higher latitudes. To a great extent, therefore, the weather and climate are home-made and not imported products, as they largely are over most of the earth.

That is, during the season, long or short, that the land is cooler than the ocean there is a more or less continuous drainage wind out to sea from the vast reservoir of cool air over the entire basin, a reservoir kept full partly by local surface cooling, and partly by numerous tributary streams of mountain winds down canyons, gorges and valleys.

This season-long wind, or monsoon, being cool, does not carry a large amount of moisture, nor is it subjected to the additional cooling necessary to get out as rain any portion of such moisture as is present. Obviously not, because the surface is not warm enough to induce vigorous local convection, nor are there many advancing cold fronts to forcefully push up the air in their way—up to where it will be sufficiently cooled by expansion to have to part with a portion of whatever supply of moisture it had before its turbulent eviction from the lower and warmer levels.

Finally, as the general flow of this air is from cooler to warmer areas, there can be but little of that drizzle precipitation that falls from very low clouds when cooled by radiation only, and not by ascent however caused, orographically (by a mountain across the wind) or thermally.

With the oncoming of the warmer season the land heats up much faster than the ocean surface and soon becomes the warmer of the two whereupon, and during the whole of that season, the wind blows back from land and brings with it abundant moisture, being itself warm and having come a long ways over a warm ocean.

And from this great and season-long supply of moisture there is condensed in turn a vast amount of rain, heavy wherever these winds flow up over a mountain range, as they do over the Western Ghats that run nearly north and south, not far from Bombay; heavy also locally incident to heat thunderstorms; and heaviest of all where the entrapped air rises in greatest amount along the enclosing mountain walls. Evidently, too, the transition from the cool dry season to that of higher temperature and copious rain can not be abrupt, but by more or less gradual stages.

So far the deductions fit the facts, but all details necessarily are omitted since they can be deduced only from a full knowledge of the location under consideration and all the pertinent facts about its surroundings, or derived from a long series of actual observations. Of course to know what the climate of any given place really is, it is useless, when adequate actual observations are available, to deduce what it must be.

But knowledge of the climate of a place is most useful when one knows not only what it is but also why it is. Here, as elsewhere, the bare facts of observation and experiment may be all that is needed in the bread-and-butter portion of our existence, but mental peace and pleasure come only when we can give a rational answer to a reasonable "why."

But it is not so easy in every case as in those discussed above, nor perhaps, in most cases, to deduce what the approximate climate of a given place, or region, even, must be and therefore is. It is pretty obvious that on a steep snow-covered slope, with an extensive higher snow covered area back of it, there must be almost constant drainage winds that sometimes reach hurricane velocity.

And that is what Sir Douglas Mawson did find on a typical place of this kind during his stay in Antarctica. One may feel reasonably certain also that an extensive land area in high latitudes must have cold winters. That fits northern Canada.

Presumably, too, the more completely such a region is prevented by high mountains, or other land barriers, from exchanging with tropical areas frigid air for warm, the bleaker its winters must be, just as they are (coldest on the earth) in central to eastern Siberia from which warm winds are shut out, and where the icy air is hemmed in until it overflows via China, largely, giving it a 10° to 15° colder winter, latitude for latitude, than that of the United States. Not only for such places as these, but for many others as well, it is more or less obvious what the general climatic features must be, but for all that there are many surprises for any one who goes, even a little ways, into the details.

Certainly no one with even a smattering knowledge of meteorology would suspect that any place near the equator can have a small annual amount of rainfall, and assuredly not if it happens also to be an island in mid-ocean, and yet Malden island, near the equator in the mid Pacific has recorded less than 4 inches in an entire year. It also has recorded nearly 100 inches in the same length of time.

The 100 inches of rainfall in a year on a tropical island in mid-ocean is not much of a surprise, but a fall of only 4 inches in a year at such a place is astonishing—a little desert in the midst of an ocean of warm water.

Certainly, too, no one who hadn't been told would say that in the course of a year Central Alaska has temperatures of 90°, or higher, in the shade on more days than has Panama, and yet, owing to the proximity of Panama to the ocean, and the great length of the summer day in Alaska, this apparently impossible climatic contrast between these two places does actually exist.

Again, who would suppose that the average January temperature at Saint Louis, on the Mississippi, possibly could be, as it actually is, the same as that of Reykjavik, Iceland? But then one of these cities is right on a course much followed by invading winds from the realms of snowshoe rabbits and polar bears while the other faces an ocean still reminiscent of the coral and the palm—a simple explanation of an incredible fact.

Here is another climatic surprise: If a Northern New Yorker should wish to spend a milder winter than his home-town affords it is certain that he would not hike to any place in Alaska, but he might, and some do, go "way down south" to Washington—Washington on the Potomac—and yet, believe it or not, its mean winter temperature is almost exactly that of Sitka, and its minimum temperatures often lower than any that ever pinch this drizzly town. The solution of this puzzle too is easy.

Many of Washington's winter winds come a long ways over land as outflows from northern snow fields where they had "frozen" long enough to get, and where they did get, severely chilled, so severely, indeed that even on reaching Washington they still may be, and frequently are, sharply reminiscent of their late conditioning, while Sitka's persistent winds, directly from contact with the still temperate Kuro Siwo are so mild that but for the everlasting drizzle drazzle winters would be just as comfortable at Sitka as they actually are at Washington. And there are lots of other surprising climatic contrasts—the world is full of them.

Rome would be ruined, at least as a winter resort, if it had the climate of Chicago, though both are on very nearly the same latitude; northern Newfoundland, where the snow lies long and deep, contrasts amazingly with the Scilly Islands where, though just as far north, even frost is so rare and so light that the wild vegetation is semitropical; and so it goes, contrast after contrast without end.

Not only are we all affected by, and therefore interested in, the present climate of whatever place happens to be our home, but also we sometimes wonder what its climate was 50 to 100 years ago, or perhaps 1,000, maybe 10,000, or even on and on through the geologic ages. For a little ways back we know the answers for many places, because we have records that give a true account however much it may differ from that which our memories provide—provide from fading recollections of a mere few outstanding occasions.

If, however, we want to know what the climate of this or that particular place was 1,000 years ago, we can get only a general idea of it: By inference from the location and environment of the place in question; from casual comments in extant literature touching the given region; and from Nature's notations, such as growth rings of trees, deposit layers in stalactites, sediment varves, fossils, etc. Finally, if our inquiry concerns the climates of given places, or regions, millions, or hundreds of millions of years ago our only recourse for information is to the rocks themselves and the fossils they contain.

Geographers and historians have furnished a large amount of material on which "irrefutable" arguments can be based to prove that the world has, or has not, undergone profound climatic changes during the past 5,000 years, or so. Most of this evidence pertains to the general region of the Mediterranean Sea, but the fact that this is a relatively limited area does not greatly matter since any distinct and persistent change of its climate certainly would be accompanied by appropriate changes elsewhere.

Much of the most impressive evidence that during the last 2,500 years the climate in this region has changed, is furnished by North Africa. Here are ruins of cities, aqueducts, oil presses, and abundant other evidences that prove this particular area once supported a population far more numerous than that which now lives there, and history tells us that this region furnished vast supplies of grain and olive oil to Rome in the years of her grandeur.

Clearly, then, so many argue, this once fruitful region could not have become barren and waste other than as a result of a radical change in its climate—at least a reduction of its annual rainfall to but a fraction of its value in the days, 2,000 years ago, when all the world paid tribute to Rome. But this gloomy inference is not accepted by all students of this subject.

Some who have given to it much and careful attention hold that deforestation, wastage of the soil, neglect to utilize to best advantage the rainfall that does occur, and the keeping by the few lingering inhabitants of a kind of goat that eats everything but gravels, have been quite sufficient of themselves, without any change of climate to make barren that or any other region.

And this hopeful surmise is being developed into a conviction by French engineers and agriculturalists who find that repaired ancient reservoirs again fill with water; that long abandoned irrigation ditches, when fully reconditioned, again, as obviously they did in olden times, can and do carry to field and orchard enough water for their needs, and that vast olive groves can now thrive where without this conservation of the rainfall none could, and none did.

The evidence gotten from North Africa, however, is not all we have bearing on the problem of climatic changes during historic times. Many invasions of one country by the inhabitants of another have been attributed to hunger incident to deterioration in the home climate. Perhaps that was the cause in some cases, but lust for loot seems occasionally to have been the compelling incentive, and sometimes, too, just a murderous craze for conquest.

At any rate, the invasion of a peaceful country by hordes from a less prosperous one does not always imply compulsion by a deterioration of climate. Hence human evidence of this sort generally needs to be checked by records kept by Nature. Here, too, one must be cautious for she never stops to annotate what she has written. Often, though, she tells the same thing in different ways so that by inter checking them, each against all the others, a closer approach to the truth generally is obtained than can be gotten from any single one alone.

One of Nature's most interesting climatic records is in the form of rings of tree-growth. A few trees, especially the older of the big trees of California, have records of this kind that run back 3,000 years, or more. However, so many things affect the rate of growth of any particular tree, and the thickness of its rings, that interpretations from them of what the climate of a given period was are difficult to make, and of themselves inconclusive.

This is obvious from the fact that many adjacent rings differ greatly in thickness when the instrumental records, kept at the same time and place show that the climate, as we count it, had not changed at all.

Obviously, if the rings of similarly located trees of the same species should be of substantially the same thickness during a long period of years, and then change to a distinctly different thickness through another long period, that fact could properly be regarded as strong evidence that at least the local climate also had changed at the same time, but the run of the rings is not so simple and regular as that, but, on the contrary, so irregular that their interpretation in terms of climate is neither simple nor certain.

Somewhat analogous to annual tree rings are annual silt deposits, varves, in lakes and other suitable bodies of water. Obviously there are flood varves as well as annual varves, but it is easy to distinguish between them.

Evidently, then, to know how long it took an old lake to silt up it is only necessary to cut down through it and count the number of these annual layers, a process that is quite the same as getting the age of a tree by counting its annual growth rings. Clearly, too, as the thickness of the silt deposits varies from year to year it is easy to match the varves of one lake bed with those of another of the same general region, and thereby go chronologically, from a lake still filling, back step by step through recently filled beds to older and still older, so long as suitable material may be available, and know all through the series, however long or short, exactly when each particular varve was deposited.

In this way the Swedish geologist de Geer found that the retreating front of the last great ice sheet was at Stockholm about 9,000 years ago. Since that time this front has continued its retreat, as, in general, have all glacier fronts, until it now is far back along the higher mountains. And as it moved north so also must have the average track of the cyclonic, or rain bearing, storms in their course across Europe —must have, since such storms normally travel along the border between relatively warm and comparatively cool air, or over a route across which the temperature changes most rapidly with distance. That is why such storms come, on the average, to lower latitudes in winter than in summer.

These facts now require a further consideration of the problem of climatic changes in Northern Africa. From the above considerations it seems reasonably certain that formerly more traveling cyclones, more carriers of general rains, must have gone along North Africa, along the Mediterranean Sea, and on beyond, than now visit these regions, and have then brought them more rain than they now get.

Hence, since the great glacial fronts have been retreating during the past 2,000 years, just as they had been since the time of their greatest extent, some 25,000 years ago, it follows that the annual rainfall of North Africa, the region under consideration, and evidently of many other places as well, must be somewhat less now than it was in the days of the Roman Empire.

Nevertheless, it may still be true, and doubtless is, that the major factors in the deterioration of the once fruitful orchards and fields between the Mediterranean Sea and the Sahara Desert were neglect of soil conservation, waste of available water and the introduction of the pantophagic goat.

No matter though how scanty and indecisive the evidence of climatic changes at this or that particular locality during the last 1,000 years, or even 5,000 years, the certainty remains that ever since, broadly speaking, the great ice sheets had their maximum extent, the climate of every place has changed, profoundly, of course, all over the ice-abandoned areas; markedly for some distance beyond their outermost borders, and least near the equator where the weather, and therefore the climate, is essentially a local product.

But if this greatest extent of the last ice sheets occurred only 25,000 years ago, plus or minus, as we say to avoid making an error, surely there was an earlier time, and a still earlier, as far back as we may wish to go, and curiosity impels us to ask what the climates of the earth were through all these ancient ages—curiosity, as definitely the mother of discovery as necessity is of invention.

In great measure, too, curiosity has been satisfied in this particular, at least as to the major features of these climates, by that shrewd and ever prying sherlock of the rocks, the geologist. He tells us that since as far back as he yet has deciphered Nature's petroglyphs, once in every 250 million years, more or less, continents and mountains have risen in relatively short times to unwonted heights, and then eroded to long-enduring low levels and much shrunken areas; and that with every great rise and spread of the land there came lower temperatures-cooler weather and colder climates—followed by increasing temperatures and warming climates as the mountains and plateaus washed away and the seas spread inland.

At each cool period there were millions of square miles of glaciers, 12 million some 25,000 years ago, the time of their last maximum extent, but little to no permanent ice at all during the warm periods when the continents were low and the pole to equator circulation of the oceans largely unrestricted. Just now the condition of the earth is about midway between maximum glaciation and its other extreme of no permanent ice, even at the poles.

At present the Wyville Thomson ridge that runs from north Scotland to Iceland, is far enough below the surface to give free access of the North Atlantic drift to the Norwegian Sea, and to the Arctic Ocean on beyond. Hence not only the west coast of Norway, but even the coasts of Sweden and Finland, beyond the Arctic Circle, are ice free, or nearly so, the year round.

If, however, the crest of this ridge were above the ocean surface, as many good geologists hold that at times it has been, then evidently the Norwegian Sea, deprived of warm water from any source, would be frozen over with a deep sheet of ice, and all, or nearly all, of Scandinavia, at least, as definitely covered with glacial ice as now is Greenland, of no higher latitude, and actually not quite as completely deprived of the tempering influence of warm water as Scandinavia then would be.

At the same time, owing to the deflection southward of a larger portion of the Atlantic drift, Ireland, England and the northwest coast of continental Europe all surely would become somewhat warmer, especially during the winter, and rainier. Presumably, therefore, it is quite possible for the climate of one place to become colder while that of even a neighboring place is becoming warmer, and both for the same reason, such as the shifting of a warm ocean current.

There also are great ocean valves, as well as dams, that have opened and closed at least once, and probably oftener, during the geologic past, with, in each case, marked effects on the climates of extensive regions. One of these is the Being Strait through which there is a vast outflow of ice water from the Arctic Ocean that necessarily decreases the temperature of the Kuro Siwo in the region of the Aleutian Islands and on down the west coast of North America as far as it goes.

If the crest of the Wyville Thomson ridge were above the surface at the same time that Alaska and Siberia were land-connected, clearly then the Arctic Ocean would be practically cut off from both the Atlantic and the Pacific. It not only would be deprived of their heat, so far as that is distributed by water circulation, but also have to keep its own cold until it covered itself at every point with a thick skin of ice.

So far as exchange of cold for heat is concerned the arctic regions then would be in the same condition is the present Antarctic continent. The cold arctic would become colder than before and the relatively warm north Pacific and north Atlantic warmer still. Again if there were a wide channel, such as long ago there was, between the Caribbean Sea and the Pacific Ocean through which a large part of the warm water that now flows northward in the Atlantic Ocean then could, and would, go on into the Pacific Ocean, it would strengthen the warm Kuro Siwo and make still warmer all the regions it affects.

Then, too, there was a time when a great sea extended all the way from the Gulf of Mexico to the Arctic Ocean, along the present Mississippi and Mackenzie valleys. This vast connecting sea between polar and tropical waters, if reestablished now, clearly would, in different ways, alter the climates of many places: Through an increased rate of exchange between the cold water of one latitude and the warm water of another; through an increase in the humidity and precipitation all along, but especially to the eastward of, this axis of North America; through giving a marine climate to extensive areas that now are afflicted with great temperature extremes; and in other ways less obvious.

Also, in the geologic past there was at one time, perhaps at several different times, a broad connection between the Indian Ocean and the Arctic Ocean by way of the Persian Gulf, the Caspian Sea, and Siberian valleys. That such a connection across Asia between the Arctic and the Indian oceans would, if reestablished, produce marked climatic effects all along its course, and even far to the eastward, is quite obvious, as is also the general type or nature of these effects.

But none of these great geologic events, the lifting of the Wyville Thomson ridge above the water surface; the connecting of the Arctic Ocean with the Gulf of Mexico through a vast North American inland sea, and with the Indian Ocean by another great sea, of which the Caspian is the lone remnant, across central Asia; the opening and closing of the barrier between the Caribbean and the Pacific; nor any other probably ever occurred alone while the rest of the world remained just as it was before, but simultaneously with many others of greater or less importance the world over.

Obviously so whenever, no matter how caused, the oceans are drained, as in the last glacial period they were, of some 50 thousand million tons of water, to pile up as ice on the land, for such a shift would lower by 450 feet the level of every present sea, expand the continents, contract the oceans and cramp their circulation, each of which effects, and others besides, would but make bad matters worse—would further chill the already cold earth. Evidently, too, with the melting of this vast volume of ice the oceans are flooded and all the effects just described exactly reversed.

This building up and melting away of the great ice sheets has run its cyclic course several times, geologists tell us, in the last 2 or 3 thousand thousand years; and this ice presumably will go again and come again at ages-long intervals, world without end—almost—as it came and went in the aeons forby. However, even if this longest range forecast is approximately accurate, it does not much concern, in a practical way, people now living, nor will it concern their descendents even of the third or fourth generation.

What, then, will be the climates of the immediate future, and of the near future? Will they change, and if so in what sense and to what extent, or will they remain many years as they now are? Judging from the past, in this case our only, but excellent, prophet of the future, it seems extremely likely that for another millennium, at least, the climates of nearly all parts of the world will continue to be substantially what they are today.

Presumably they will not run smoothly, they never have, but fluctuate, year by year, and even have trends in the same sense, or direction, through several to many years with serious results, trends that will come to an end, as heretofore, and be followed by a reversion to normal, or near normal, conditions. These, however, will not be constant conditions, but, on the contrary, a very slow warming up of practically every region outside the Tropics, incident to the continuation of the slow retreat of all great glaciers, a retreat that has been going on for many centuries, and that is now going on at a decidedly measurable rate.

Indeed the few fairly reliable century-long temperature records that are available seem to show that the higher latitude areas are now a trifle warmer than they were a hundred years ago, just as we would expect them to be from the known shrinkage in that time of numerous glaciers in many parts of the world.

This gradual melting of the ice sheets, great and small, and progressive slow increase of the average temperature of the world as a whole, and of the higher latitude regions especially, we confidently expect, because just that has occurred several times in the geologic past; because it is now in active progress, and has been for many centuries; and because we know of nothing whatever that promises to reverse, or even slow up, this route of the glaciers.

After a few thousand years, 10 perhaps, the polar and adjacent regions will have become so much warmer than they now are that the temperature at either pole and that at the equator will differ far less than they do at present.

Hence the general circulation of the atmosphere, driven by the pressure differences, level for level, incident to temperature contrasts, will be correspondingly feebler; evaporation, facilitated by air movement, will be less; rainfall, dependent on evaporation, will become more and more scanty; the cyclonic storms of the middle latitudes will decrease in frequency and average intensity, and run their courses in higher latitudes.

Hence the arid and semiarid regions will so broaden as to bring desolation to much of the world's present most populous areas. The tropical regions will still abound in thunderstorms, much as they do at present, and the high-latitude areas in drizzles by summer and the settling of fine snow in the winter. By and by, in terms of millions of years, the mountains will wear down, the plateaus shrink to lower levels, the oceans flood all valleys and drown vast areas that now are populous and fruitful lands.

Drab and dreary the world will be, with all climate and little weather. Nor change will there be through millennium after millennium until the world again crumples up mountains-how and why we do not fully know—builds plateaus, and the oceans give abundantly of their waters to make ice upon the lands. Then the earth again will have its brief season of vigorous winter sports followed when tired out, by another long summer siesta. And after that what? Probably much the same things over and over again as long as sea and land shall endure.

Chapter 5

Weather Control

Fussing about the weather comes as natural to a farmer as cussin' to a sailor, and there is good reason for it too. If his crop is corn, either too much rain or too little keeps him from getting it planted in proper time, and so do unseasonably low temperatures, even when the rains are just right.

Again a heavy frost after the corn is up, such as occasionally occurs, always is very injurious. After this danger is safely passed, if indeed it is, there may be so much rain that cultivation, necessary for vigorous growth, is impossible; or the ground may be so dry that cultivation would do more harm than good.

Finally, even if all goes well until autumn is near and a bumper crop all but assured, an early killing frost may occur that over night turns the pleasure of anticipated profits into the sadness of an assured loss. Or, if his crop be wheat, or something else, instead of corn, his weather worries, though somewhat different, are none the less serious.

Of course, then, the farmer is a weather grouch, nor can we ask him to be otherwise so long as he is human. Indeed we who are not farmers would grouch worse than he does if Nature were all the time mocking us as she mocks him.

Nor is the farmer the only one whose success or failure is determined by the weather. The sailor, for instance, hoping for favorable winds may instead be becalmed for weeks, or wrecked by a storm, and therefore he, too, has many an occasion to express his sentiments in sonorous terms of certain meaning.

And so have all the rest of us, refrain as we may, for none is wholly exempt from weather worries since feasts and festivals, celebrations and games, and everything else, from picnics to pitched battles, that take us outdoors are affected by the weather.

A desire, therefore, to do something about the weather, or at least to have something done about it, was one of our earliest inspirations. In the beginning we were not wise enough to be foolish, and so instead of trying to boss the weather ourselves, or get some other power to mould it to our personal liking, we only sought protection from its disagreeable elements, which protection we obtained, and, on emergency, still obtain, by going into some natural shelter hollow tree to empty cave.

So convenient and so efficient was the cave as a protection against heat and cold, rain, snow, blow and bluster, that men soon began burrowing artificial caves into the sides of cliffs for the same sort of protection, or down into the ground, and in portions of the world still do so to their own safety and comfort.

Then, ages after holes first were cut into cliffs for cave-like homes, some super genius conceived the revolutionary idea of leaving the "hole" alone and building an artificial cliff around it, and low and behold there came to pass the walled house, that still is evolving with every change in the kind and quality of structural material available. However, this evolution has not been all-to-the-good in every particular and for all things; indeed in some cases improvement has been obtained by reversion to the more primitive.

Theaters, for example, no longer need holes in their walls for windows, only one big opening for entrance and exit; and some factories are constructed in the same primitive way, though, of course, using artificial light and conditioned air. The next logical step obviously would be to bore into cliffs, or burrow into the ground, just as did the troglodites of old, and there with the turn of a knob regulate the weather to our own liking, and let the winds above go whistle.

This may seem to be a terribly distorted glimpse into the future, but really its sole error has but to do with tense—instead of saying that the next logical step *would* be, to be exact we should say *has* been, for already there are busy underground stores of all sorts, barber shops, restaurants, and all the other things essential to a continued existence of safety and comfort—no freezing wind nor sweltering beat; no pelting hail nor driving rain; no icy pavements to break our legs on nor jagannath car to bump us into heaven before we are half ready.

No, this is not that paradise all of us hopefully anticipate, but just a well managed troglopolis, such as thousands of us see every day along the underground approach to the Grand Central Station in New York City, and in the subways of Paris, London and other great and crowded cities. Here the weather not only is controlled, but made as well, just as it is in theaters, department stores, offices, homes, hospitals, railroad coaches and whatsoever other enclosure from chicken coops to royal palaces.

Another ancient and successful effort to control our own personal weather was the adoption of appropriate body covering, from hardly "nothing at all" in tropical lowlands to the multifold furs for Arctic winters.

But however obvious and important clothing, caves and houses may be as controllers of one's personal weather, they do not afford the kind of general, outdoor regulation we commonly have in mind when talking on that subject. We can do what we like, as just explained, with indoor, or domestic weather, but only a little with the wild weather of forest and field.

Nevertheless even that little is well worth while. Flooding often is effective in protecting cranberries from destruction by frost and freeze; many a valuable crop, vegetable, apple and citrus, has been saved from ruin by properly distributed small fires; and a cheesecloth canopy may afford just the degree of shade necessary to perfection in certain valuable kinds of tobacco that otherwise would be poor and profitless.

Wind breaks, also, whether consisting of a row of suitable trees or just a plank fence, are helpful in mitigating certain wind effects, such as making unmarketable windfalls of otherwise first-class fruit, and the piling of impassable snow-drifts onto highways and railroads. In a few cases this winter nuisance has been avoided by the building of mile-long houses over badly exposed sections of a railway—expensive, of course, but better, even so, than a complete blocking of traffic.

All the above devices, holes to houses and furs to fires, are means for weather dodging, and not in any sense methods of controlling the state and condition of the outdoors atmosphere over any considerable area, the result that through the ages many have desired, some have honestly tried to produce and a few have claimed the power to do in any manner desired— to deliver whatever sort of weather to any place at any time-and he finds "suckers," too.

In Java, it is said, a weather doctor sometimes was employed to "prop up the clouds" (prevent during all important feast, and this be would agree to do, for a consideration, on condition that so long as the celebration might last neither the rich man giving the feast, nor any member of his family, nor any one of his host of servants, would wash any clothing or even indulge in the luxury of a bath, and that every one of them, male and female, would, during all that time, observe strict chastity—conditions difficult to meet in this tropical island whose natives are passionately devoted to the bath, and well calculated to furnish a convenient alibi likely to be needed, under the circumstances, in a country of such frequent rains.

Of course, all this is just funny to us who, in the pride of our intellect, boast that "nothing like that can happen here." But it call and does happen here, wherever "here" may be. Only a few years ago a weather wizard was paid, according to reports, $1,000 per day to prevent the occurrence of rain during a fashionable horse show; and shortly thereafter engaged to bring rain over a particular dry region, not for the sake of parched pastures, but to get the fields forests in such condition that a running fox would leave a trail behind him that hounds could follow.

As in these cases so also in most others, rain is the weather element that we most need, or think we need, to bring under human guidance and direction, and naturally so because it is vital not only to our comfort but even to our very existence.

The earliest attempts to obtain rain evidently were by forms of magic, especially of the imitative type, or sort of showing by example of the kind of performance desired, in the expectation that this would lead to the like actions by the clouds, or by the vaguely conceived being of some sort that controls the clouds. Such an idea is not without logic, granted that rain and shine are but responses to the will of a sentient being kindly disposed to please whenever informed of our needs.

It is abundantly supported also by many analogous practices. Thus when we want a monkey to do tricks we teach him by example, complete or partial; and the child too learns in the same way before he can talk; nor indeed does he, or any of us, ever become wholly immune to the influence of suggestion, or the tendency to unconsciously imitate.

Therefore, reasoned the primitive mind of the old stone age, and reasons the untutored mind of today, if we want the "rain god" to send down refreshing showers we must indicate to him in a syncopated manner what our desires are—show him by rhythmically repeated example just what we want him to do, and presently, like the monkey or the child, he will imitate our actions, and, whether he knows it or not, be induced, or tricked, into doing just what we want done.

One of the most impressive and realistic of these bits of imitative magic was still practiced not long ago, we are told, in Esthonia where, to induce the clouds to give a sorely needed rain, three men climbed fir trees in a sacred grove and jointly gave a dramatic and unmistakable imitation of exactly what was wanted.

One carried fire brands which, when knocked together on a dark night, furnished, without the aid of much imagination, an excellent imitation of lightning; another beat on a kettle, or some other suitable vessel, to simulate well the roll of thunder that follows every brilliant flash; while the third, most realistic of all, poured out an appropriate shower from a watering can.

Asleep, indeed, Nature must have been any time she failed to see this beautiful reproduction of one of her finest dramas; stupid, if, seeing she did not sense its meaning; callous, incredibly, if seeing and understanding she yet gave no heed to the hunger and thirst of man and beast. Sometimes the rains came—they were on the way anyhow-and sometimes they did not, for Nature, being neither man nor monkey, is wholly without psychological response.

The most widely practiced rain-making magic was, and is, some form or other of getting wet, or, more frequently, ducking or drenching the other fellow. In central Celebes, where summer is perpetual, bathing a passion, and getting wet a pleasure, a party of young folks go to a convenient stream, when a shower is wanted, and there splash each other until all are as wet as any rain could make them.

In Serbia the magic to bring rain was to take a girl as the star performer, strip her of all clothing and then cover her from head to foot with flowers, grass, and other things to which rain is essential, and march her through the streets at the head of a troop of singing girls and, at every dwelling, have her soused by the credulous housewife with a pail of water.

The records are not at hand from which the efficacy of this method of breaking a drought might be inferred, but it is quite certain that the "flower girl" was glad enough when the wet spell was over.

Another, but cruder, form of this getting-wet magic to bring rain was tried in Rumania, according, it is said, to official reports. In this case the drought was so severe that the peasants were ready to resort to the most heroic methods to break it, and believing that for that purpose a bath was efficacious, but being afraid themselves to risk the dangers of this wholly strange act, got their wives together and compelled them to bathe. Even a drought may now and then do a little good!

In southern Russia the wetting magic for rain sometimes was mixed up, we are told, with a crude religious element. Instead of just pitching a stranger into a river, as the women occasionally did, the peasants would grab a priest, immediately following the service, pin him to the floor, and there give him, robes and all, a thorough drenching. In this way the vicar, at least, was made to understand very clearly just what was wanted, and what he might expect if it was not soon provided.

Of course, though, some husky priest might fail to recognize meek submission to such treatment as a part of his pious duty and resent it accordingly, and so the Armenians, at least, took the safer course of drenching the priest's wife, which no doubt was just as effective.

In some places the most approved rain-bringing magic was the simulation of a cloud with something dark, such as a black horse or black cow. One procedure with the horse was to face him west and rub him with a piece of black cloth until he neighed. What role the neighing played in this curious bit of magic is not clear, but it is certain that to the drought distressed it was a most cheering assurance of speedy relief.

In Queens land the natives to inspire precipitation gather around a small artificial pond, and there some of them, by flapping their arms and quacking, pretend to be ducks, while others hop about and croak like frogs—dramatic indeed, and full of hope for the faithful, however indifferent great Jupiter Pluvius might remain.

These are but typical examples of the endlessly varied forms of rainmaking magic. And as often it rains more than we want, or at an inconvenient time, there developed also different forms of rain-dispelling magic.

One was to point a firebrand in the direction of the rising sun and then slowly so turn around with it that it would follow the normal course of the sun across the heavens, saying by pantomime, supplemented, perhaps, by loud voice as well: "That's what we want you to do, Oh sun, shine now from morning to night, for already there has been too much rain."

Another piece of rain dispelling magic is to put outdoors hot stones that evaporate the rain as fast as it falls on them, a result so discouraging to the rainmaking spirit that presently it gives up trying to make things wet. Indeed undoing anything the rain had done, such as drying one's clothing or other wet objects, has been, and by some may still be, held very effective in bringing fair weather.

But efforts to control the weather, especially efforts to induce rain, have not been, and are not, all just so much magic. Many of them have religious elements in the sense of appeals to departed ancestors, traditionally great medicine men, or some vaguely conceived supernatural power.

In some cases food as a persuasive offering for a needed rain is spread out before a cherished ancestral skull, or row of such skulls. Another magico-religious procedure is to pour water on the grave of some former rainmaker whose reputation has grown mightily in the rich soil of tradition and thus beseech him to come again, as of old, to the rescue of the hungered and the thirsty.

In some places, where the rod is freely used and compulsion is the rule, drastic measures occasionally are taken to compel whoso supposedly can to forthwith bring the needed rain. The simplest of these methods is to throw rocks at the grave of a departed master rainmaker—practiced, not of course, but actually, only by barbarians. Not of course, advisedly, because a custom, sillier, if possible, gruesome and revolting, once obtained, and still lingers, at least in belief, it is said, in regions that proudly claim to be intelligent and civilized, namely, the custom of digging up the dead, especially friendless criminals, and dumping them into a swamp, when rain is greatly needed, or of throwing their beads, at least, in a river.

Sometimes a logical savage, in the belief that a particular kind of animal, usually a frog, is an especial favourite of the rain deity, and being unable to influence the god directly, tries to get the desired by cleverly subjecting that pet to some condition in which it must have rain or perish; most of which rain, the persecutor believes, incidentally will fall where it will be of personal benefit to himself.

Sometimes, and this is not confined to the "barbarian," dire threats are voiced as to what will be done to an image of, or to something dedicated to, the god of the weather if rain does not quickly come to revive the dying vegetation and perishing cattle.

Not only are threats made, but even insults and abuses often are heaped upon him who supposedly could, but actually does not, procure the wanted rain. In India, to break a drought, pictures of the rain god Megha Raja, are painted heels up, head down, and placed out in the sunshine until disgrace and discomfort bring him to the performance of his duties.

Ah well, that bit of absurdity, we say complacently, occurs in far-off India—nothing like it here. But is it any more, or less, absurd than ducking an image of Saint Peter in a river when prayers for rain appear to go unheeded? And is it not cut to the same pattern of rationality as standing Saint Joseph on his head when long waiting for a wish has sorely tried one's patience? When it comes to absurd notions we don't have to go far to find some one living in a very thin walled glass house, and who therefore should be exceedingly cautious about throwing stones.

No less absurd, we now believe and dare to say, were the various weather talismans common in Europe during the early and middle ages, and indeed still lingering in places. Perhaps the most numerous of these were the so-called hail crosses, erected in the belief that they ward off hail and thereby protect the grape, especially, and other crops from serious damage.

Another widely dispersed religious talisman was the church bell, blessed at the time of its casting, bearing an appropriate motto, and often rung to calm the turbulent elements at the time of a severe thunderstorm. The child-like faith that so many once had, and that some still have, in these and other talismans, and charms, deserves our highest respect; nevertheless honest deduction of effect from cause and induction of cause from effect, rigidly tested with observation and experiment, have shifted our faith.

Few now in any well informed community believe in the efficacy of magic, of whatever form, as a means of controlling the weather. They hold, too, that the course of the weather is not in the least altered by whipping a frog, by hanging up a dead blacksnake, by stoning graves, by placing food before grinning skulls, nor by any other sort of threat, abuse or supplication.

However, as these ways of trying to influence the weather, especially to the point of bringing an abundant rain, gradually were abandoned, others as ineffectual, though generally based on sound scientific principles, were advocated and a few of them even tested by ample experiments.

It is sound practice, as everyone knows, to bring indoors on a cold night a plant or other object that frost would injure, or to put over it a suitable covering, that is, a woolen, or other loosely woven, fabric; but not a metal vessel, such as a tin bucket, as that would just make matters worse.

Also, when the wind is light to calm, frost may be kept from an orchard, say, by an adequate number of properly distributed heating devices, such as the widely used burner of crude oil. But these obvious and practical weather controls have nothing to do with the production of rain, the result more generally desired and for which every severe drought grows a luxurious but useless crop of devices, nearly all old to ancient and long discredited.

It was noted even in ancient times, and again and again down to the present, that usually a big battle soon is followed by a rain. This observed fact, for it is a fact, was variously explained. Some argued that the gods abhor the sight of a field of carnage and send heavy rains as soon as possible to wash it off.

Others held that the rains are the condensed vapors that arise from hot blood and profuse perspiration. Still others argued, especially after the use of cannons, that the rain is caused by the jostling together of cloud droplets by the great noise of many batteries.

This reasoning appeared so plausible that at the time of every severe drought many people urged the liberal firing of cannon, the exploding of dynamite, or anything else that would make a loud noise, as a sure means of quickly bringing a good rain.

It was tried, too, several times, but always without the expected results. In fact, as we now know, cloud droplets are not appreciably jostled together by sound waves. Besides, it seems quite unlikely that the droplets commonly would coalesce even if bumped together. Anyhow, no amount of noise made for the purpose has ever yet brought encouraging results. But, as stated above, rain commonly does soon follow a big battle, especially one fought in the open.

The reason is simple: Such battles are fought mainly in regions where rain occurs about once in 5 to 10 days, and usually are forced by the attacking party only when the ground is hard enough to facilitate the movement of artillery, that is, not until at least two or three days after a good rain. If, then, the battle itself should last, as it well might, a couple of days, the next rain normally due cannot be far off. No, battles have nothing to do with rains, but rains have much to do with battles.

Another plan for getting rain, and in this case a theoretically perfect one, was independently devised, by many an inventive genius, immediately after it became widely known that when the temperature of a particular body of air is steadily decreased presently some of its moisture begins to condense out, and that ascending air, however caused to rise, grows colder and colder with increase of height.

Clearly, then, a good rain can be had any time at any place by just causing an adequate amount of humid air to rise there high enough—a couple of miles, say. But how to get so much air up to such a height is an unsolved problem that, from the very nature of things, well understood, is going to remain unsolved.

One obvious plan, often proposed, for forcing the air up to great heights, is to form a ring of many airplanes with their noses down and out, like muskoxen on the defence, and set their propellers going at full speed. This plan looks good on first sight, but a little calculation, based on what we know of the atmosphere, proves that, even if any rain could be gotten in this way, it would cost far more than importing the same amount of water a thousand miles by tank.

Well, then, others have said, just build a high mountain across the path of the wind so as to deflect it upwards to the necessary height. A fine plan, this, so far as getting some rain from the passing winds is concerned when the state of the air is favorable, but of course utterly impracticable.

Some of the more scientifically minded suggested that the air be made to ascend by heating it with a big fire. Academically this is an excellent plan, for sometimes the air above a raging forest fire does rise high enough to produce a cloud from which, on rare occasions, even a little rain may fall; but this method of getting a light shower, once in many times, perhaps, is too much like burning the barn to get rid of the rats ever to be intentionally adopted.

The rainmaking plan most frequently urged, occasionally tried, and that at first seems safe, sane and certain, is to put something way up in the free air that will cool it; cracked ice thrown out from balloons, was the first method suggested, but now the usual proposition is to empty out gallons of liquid air from a plane sent up for that purpose.

This interesting plan works—on a small scale, of course, for the amount of liquid air available necessarily is relatively small; but astonishingly enough it only works backwards. That is, the chilled air becomes denser, through contraction, than that surrounding it, and therefore sinks to lower levels where it becomes warmer and drier than it was before it was first cooled.

Liquid air, if sprinkled in considerable amount on a small cloud, would cause it to evaporate, and experiment has produced just that result, but rain has not been induced in this way, and evidently cannot be, since it is actually a warming and drying operation however paradoxical that may seem to be.

A wholly different scientific scheme of rainmaking is based on the experimental fact that cloud forms readily only in air that contains innumerable condensation nuclei, such as microscopic particles of salt or other hygroscopic substance. That was hint enough for the rainmaker. At once he insisted that bushels of dust be scattered in the sky from planes to give the moisture something on which it could condense, and on which, in his opinion, it promptly would condense and quickly produce a rain.

This plan also has been fully tested experimentally. It did not work, and for the very good reason that the air always contains more nuclei than ever are needed, so that adding more cannot and does not do any good.

The latest really scientific effort to produce rain was that of scattering electrified sand on the tops, or upper surfaces, of clouds in the expectation that each charged grain would gather unto itself, by attraction and collision, enough particles to make a good sized rain drop by the time it got down through the cloud.

It was patiently tried and tried, but there came no rain, at least none that with certainty could be attributed to the electrified sand. Furthermore, it has been shown by a mathematical discussion of the subject, based on the known forces involved, that not enough cloud droplets can be captured in this way to produce a noticeable shower.

Scientific methods also have been employed to dissipate hail storms, and even to prevent entirely their occurrence over a particular region. The method of mitigation, not long ago extensively used, and still practiced to some extent, was, and is, to shoot vortex rings of smoke against the thunderstorm cloud with flaring, or funnel-shaped, hail cannon. Of course, though, the storms went on "regardless," as any physicist could have assured the wasters of ammunition that they would.

Prevention, more desirable, of course, than mitigation, was attempted on a large scale a few decades ago by erecting in vineyards, and elsewhere in the open, giant lightning rods, from the tops of which brush discharges streamed off at the near approach of a hail storm.

The idea that hail would thus be prevented was based on the belief that it is caused by the electrification of the thunder cloud, and the further notion that this electrification would be dissipated, or neutralized by the slight coronal, or brush, discharges from the pointed tips of the vertical metal rods. Well, the storms came, the winds blew, the lightning flashed and hail fell the same as before. Of course, because the electrical condition of a thunder cloud has nothing to do with the formation of its hail; and even if it had, the feeble coronal discharges could not appreciably affect that condition.

No sooner, however, are efforts to control the weather blocked in one place than they break out in another. The latest, not yet quite given up, is the highly desirable one of clearing harbors and aviation fields, especially, of fog.

Certainly fog can be cleared away at any time from a small space; by high-tension electricity, as it was generations ago; by heating the fog and, incidentally, the air it is in, as that changes the liquid droplets into invisible vapour; and by decreasing the current humidity with any substance that abundantly absorbs water vapour, and thereby induces spontaneous evaporation.

But none of these methods has yet won a place in useful service where large spaces must be cleared. In fact all efforts at weather control of whatever sort were fairly summed up by the disillusioned Italian workman on the Pacific coast when he indignantly exclaimed: "Uncle Sam one big fool; he make horn go ooh-ooh-ooh, and bell go dong-dong-dong, but the blame old fog, she just keep a coming!"

Naturally and properly enough, even those who most strongly insist that knowledge, like virtue, is its own reward, ask: "Of what practical use is meteorology anyway?"

He might readily admit that there is a real pleasure in knowing how the rainbow gets its colors, and how this and that other phenomenon of the atmosphere comes about, and yet keep coming back to the same old question: "Of what practical use is it?" And this he can, and does, ask with no thought of disparagement, but only in the laudable desire for information.

The oldest branch of "useful," or applied, meteorology, dating back thousands of years, the most used today, and most misused, is the foretelling of the coming weather for an hour to a day or two in advance, and sometimes for an appreciably longer period, with, however, a rapid approach, as that period is made greater, to the level of a sheer guess.

Judging from what people do now, and from the fact that human nature appears to remain much the same from generation to generation, we safely may assume that ever since indefinitely far back in the forgotten ages nearly every one to some extent forecasted the coming weather for himself. It is likely, too, that he then, as we now do, asked the other fellow what he thought the weather was going to be—whether he thought there would be rain in a little while, or, perhaps, the next day.

Clearly, also, some one, probably old and wrinkled, would acquire a reputation for wisdom and a foreknowledge of the coming weather and be consulted on that subject by any one and every one who had need, or curiosity, to know. This, of course, made him a quasi-official forecaster, after which his next forward step was both obvious and easy. To become an official forecaster of high repute he now needed only to get on the public pay roll.

This he appears to have done in at least one part of the world, Assyria, as long ago as the 7th century B.C., as recorded in Assurbanipal's cuneiform library, now in the British Museum. These oldest known weather forecasts seem to have been based on omens of one sort and another, including halos and coronas about the sun and the moon-"signs" of real worth on to a day or two in advance, and still much used by those who live close to Nature.

But to be told, however reliably, what the weather would be during the next few hours, or even on the morrow, did not satisfy the ancient Assyrian public any more than it satisfies the public of today. And so with the push of the populace behind them and the lure of lucre before them, these forecasters, as have many others all through the ages, foretold of floods to come, droughts, and other events of general interest, without, of course, a scintilla of rational evidence in support of their bold assertions.

But this irrationality, as we now would call it, on the part of these early forecasters, is entirely forgivable since the idea that every physical phenomenon must have had an adequate physical cause, and which we now regard as innate, or practically so, had not then even sprouted anywhere, much less taken firm root. A little later it did germinate in Greece, but its cultivation soon was neglected for the lazy practice of attributing the weather and all its changes, as also everything else of importance, to willful acts of the gods—manifestations of kindness, for which there sometimes were offerings of thanks, or evidences of anger that inspired many a pious pledge-soon forgotten.

During these long centuries the coming weather was predicted on the basis of signs—mostly worthless—or "computed" according to some confusing astrological nonsense. Near the beginning of the 16th century there began the publication of pamphlets, called "Prognostica," that foretold the weather of every day of the coming year.

Of course season and locality were taken into account, at least to the extent that heavy snow in July was not predicted for central Europe, nor a heat wave in January; but apart from this regard for the seasons, every forecast was a hit or miss that one person could make quite as well as another.

And yet, unfounded and absurd as we now know these long-range forecasts to have been, they then were regarded as well-nigh infallible. No wonder it is, therefore, that for generations Prognostica were printed and sold by the millions. After an unconscionably long time, however, a little weather gumption began to leaven the meteorological loaf, whereupon, to the same extent, faith in these predictions became so impaired, so much less childlike and implicit, that the sales of Prognostica dwindled to that level at which it was no longer profitable to publish them.

This so took the breath of life out of them that they faded away like the mist of the morn—vanished, one might have hoped, into the limbo of the unregrettably forgotten. But if Prognostica no longer might be sold at a profit, surely they still could be given away; and right there that clever opportunist, the patent medicine man, saw a golden chance and took it by flooding the country with free almanacs.

His forecast of the weather of today might be wrong here, but certainly right for some other place, and his predictions covered so wide a region that they seldom could be wrong all at once everywhere. Besides, if one should be misled and in consequence get drenched in a chilly rain, right before him was the cheering information that for only a dollar a bottle he could get at "any drug store" the renowned Redman's panacea that would cure the cold he had caught, or rheumatism, if that was his trouble, gout, toothache or anything else from baldness of pate to fallen arch.

However, this method of distributing humbug long-range forecasts flourished only a few decades and then speedily vanished, incident more to increase of legal compulsion, presumably, than to decrease of public credulity. In fact even the legal compulsion was directed wholly, or nearly so, to the falsity of the medicinal claims and not at all to the equally false claims of an allinclusive weather prescience for a year or more in advance.

However, the vanishing of the patent medicine almanac did not put an end to long-range weather forecasting, nor perhaps even greatly decrease it, but caused people to turn again to their own infallible sign's—blotches on the goose bone, fur on foxes, moss on trees, husk on corn, and dozens of others just as good!

And so it has come to pass that he who just will have a long-time forecast of the weather must either make it himself or pay someone else to furnish it for him. They appear to be equally reliable, but the first plan affords the more fun for anyone who can quit reasoning long enough to enjoy it.

This brings weather forecasting down to date along one line, and it will be interesting now to follow its evolution along its other, or rational, course. Before reasoning from cause to effect got well started, one phenomenon was just as good a sign of the coming weather as any other, but even so, long experience showed that some of them were more commonly followed within a few hours, or, at most, a day or two, with a given type of weather than were certain others, therefore selections were made.

From the most ancient times down to the present, rings around the sun and the moon have been regarded as harbingers of falling weather, not invariably, but often enough to be significant. For many centuries this was just an observed fact, but quite as valuable then as a weather sign as it is now after we have learned how these rings are produced and why they so frequently shortly precede falling weather.

Similarly, the roaring of the mountain, the agitation of the sea shore, the long carry of sound, the colour of the dawning sky, and many other natural phenomena were observed to precede shortly certain states and conditions of the weather and therefore were used, and ever since have been used, some as helpful warnings of approaching storms, and others as cheering promises of a cloudless day.

All these signs are of purely physical origin and closely associated with the movements of storms. That is why they have stood the test of time and observation. They were useful long before anything was known about their causes, nor has that knowledge decreased in any measure their value and daily use—just increased respect for and confidence in them.

In the meantime, curiosity long ago began laying the foundation for longer ranged and more reliable predictions of the coming weather than could be had from even the best of the natural or physical signs. This curiosity was to know what sort of weather they had back home, say, the time it was thus or so "here."

Doubtless innumerable questions of this sort were asked, and a great many reliable answers received, but apparently no collection and study of them has ever been made. However, a much better plan, leading to the same scientific end, was begun in 1654 under the guidance of Grand Duke Ferdinand of Tuscany. This consisted of the taking and recording of weather observations day by day at a number of widely scattered places. Unfortunately these observations were continued only a dozen years or so.

However, systematic observations of this same sort were begun again at various cooperating stations in 1780. The data thus collected for 1783 were carefully studied in 1820, by which time systematic weather observations were being taken in various places, largely for climatological purposes, especially to determine relations between climatology and agriculture, and to find the places to which persons with this or that ailment should go.

Through comparison by letter, or otherwise, of weather notes taken on special occasions, and through a study of daily records made at several different places in the same region, it soon became clear that in Europe and America, at least, storms commonly travel from west to east, not directly necessarily, but in that general course, and that, too, blow the wind whencesoever it may—even if from the east, still towards the east the storm will go.

With this fact in mind, it clearly was obvious to any one that to know what the weather of the morrow would be at any particular place, it was only necessary to know what the current weather (weather of the current day) was a day's storm-travel to the west. This was tantalizing in the extreme.

Here was a perfectly obvious and simple way of knowing exactly what the weather of the morrow, and even of the day after that, would be, and yet a way wholly impossible to use, for there were no means by which the necessary information could be sent ahead faster than the speed of the storm itself, roughly 30 miles an hour.

The invention of the telegraph, however, changed all this, and gave to whoso would forecast the coming weather new hope and a near assurance of complete success. And so it came about that within a very few years after messages began to be ticked over the first-telegraph line news of the coming weather was being obtained in various places, and forecasts based thereon publicly displayed and distributed. Storm warnings were first telegraphed by the French Government in 1855, and by the Board of Trade of England in 1861. A service of this sort for the United States was advocated by Maury, urged by Lapham, and demonstrated by Abbe, to mention only a few of the forward-looking men of their time.

So obvious were the benefits that would come from such a service that very soon, actually in 1870, it was officially established as a part of the duties of the Signal Corps of the Army. In 1890, after having grown greatly, it was transferred to the Department of Agriculture, where it continued to effect the same duties as before and to assume others as needs therefor developed. In 1940, it was again transferred, this time to the Department of Commerce—the tail, aviation, having pretty thoroughly wagged the agricultural dog. In the meantime the Army and the Navy had developed independent weather services to meet their own special needs, as had also certain airways.

Very soon after the invention of the barometer it was noticed that the height of the mercury in the glass tube usually varies slightly from hour to hour and often considerably from day to day, occasionally as much as an inch or even more.

It also was noticed that when the weather is fair the top of the mercury column generally is relatively high up in the containing glass tube, and that during foul weather it commonly stands at a lower level. It soon was noticed, too, that the level of the mercury varies gradually, obviously owing to correspondingly slow changes in the amount of air overhead, and furthermore that commonly these changes start before the weather has noticeably altered—that as a rule the level of the mercury begins to fall even before the onset of foul weather, becoming lowest about the time the storm is at its worst, and to rise before the weather has again become fair.

This was hint enough to set the clever physicist, Robert Hooke, at work on a forecasting robot. His plan was to let the rise and fall of a float on the mercury in the up-turned open end of the barometer tube move a hand over a clock-like dial, and to place on this dial appropriate weather terms, such as "very dry," where the hand stands when the barometer is near its highest; "storm," where it stands when the barometer is quite low; and "fair," "change," "rain," at appropriate intermediate positions.

Of course, it works, and while its forecasts are far from infallible, they are much better than no forecasts at all. Therefore, and partly, perhaps, because the thing seems uncanny, Hooke's "weather glass" soon became as popular as the grandfather's clock. Nevertheless, both are now outmoded, now that we can wear an aneroid barometer on one wrist and a watch on the other, save in the homes of those to whom sentiment is a vital part of their happiness, those who whate'er their station in life, still revere their childhood past.

And the old weather glass, when properly adjusted and correctly understood, still gives warnings of the approaching weather that are distinctly helpful, especially when modified, if need be, by such pertinent matters as season, location, temperature, direction and speed of the wind, kind, amount and velocity of the clouds, and anything else one's experience may have proved helpful. This clearly refers only to what can be done at any given place, if one knows how, will take the time to do it, and is supplied with the necessary instruments properly exposed.

This "if," however, is exceedingly formidable. Few are suitably equipped; few know how to interpret the composite of all their readings and observations in terms of coming weather; and few take the time to do so. Clearly, then, when all these small fractions of the whole are multiplied together, the end product is very small indeed. That is, only an occasional enthusiast takes all the necessary time and trouble to make the best forecasts possible from the readings of his own instruments at his own home.

But that is not half as bad as it seems, for now anyone, our enthusiast included, can get more reliable forecasts, and for a longer time ahead, by turning to his daily paper, listening to the radio at the right time, or, best of all when available, punching a telephone button (dialing a particular number) and hearing a polite robot say: "Fair and cooler, with light northwest winds tonight and tomorrow morning," or whatever other message may be the proper one. Wonderful, wonderful, and so much more convenient than trying to figure it all out from one's own observations! But it does make for mental atrophy.

As soon as the records of observations simultaneously made at a considerable number of places, covering a rather extensive area, were gotten together and compared several important facts were noted: That the higher the place the lower, in general, its barometer reading; that when the readings simultaneously made at the several stations are altered as nearly as possible to what they would have been if all had been made at a particular given level.

They still differ from each other, but less than they do when not so reduced; that usually these many readings fall into several groups according to their approximate values; that the stations at which each of these groups of equal readings were obtained frequently lie along a closed curve on a map of the region over which they are distributed; that the several curves, corresponding to the different reduced readings of the barometer frequently are roughly parallel to each other, or more or less concentric; that in a rainy region the height of the barometer decreases with approach to that centre; that in a region of clear skies the reading of the barometer increases with decrease of distance from this centre; that the direction and strength of the wind are definitely related to the distribution of the places whose barometers have the same corrected readings at the same time; that—through a long list of other more or less important facts and relations that have been revealed by careful studies of innumerable weather maps.

Clearly, therefore anyone minutely familiar with all these facts should be able to make at least a fairly good forecast of the morrow's weather, from a detailed map of the current state and condition of the atmosphere over the region hundreds of miles round about the place in question.

Many have tried to do so, in this purely empirical way, and a few have succeeded uncannily well. In general, these excellent few appear almost to live with their maps, and to have of them such retentive memories that as soon as a detailed map of the current weather conditions over the country is before them they commonly snap out first rate forecasts with but little effort.

Nature, however, does not act empirically, but strictly from cause to effect. Therefore, a knowledge of the causes of the many weather phenomena clearly must be helpful to the forecaster. These causes are numerous, complex and distributed throughout the atmosphere from the surface of the earth up to great heights.

Hence to forecast on the "rational" method, that is, by reasoning from cause to effect, one must have before him accurate maps of the current values of all the important weather elements at various levels. This obviously entails the speedy collection of several sorts of data, on the surface and up above, at many different places over a large area; the immediate delivery of all this raw material to a distant prognostic centre; the prompt graphical representation of all this information on suitable maps of the area covered; quick inferences by the forecaster from these maps as to what the weather will be in a few hours, and on the morrow, over this and that special region, and even at definite places.

Obviously, in this work there is no time for much deliberation or tedious balancing of the pros and cons, since in that case the weather would be past and gone before any warning of its coming could be gotten to the public.

A forecaster therefore must be a quick and experienced empiricist with the background of a detailed and accurate knowledge of the how and why of every phenomenon with which he has to deal, a knowledge that keeps his empiricism in proper bounds and guided in the right direction.

Of course, at the moment the forecaster reaches his conclusions, however much his ratiocination was conscious or subconscious, he alone has the benefit of knowing what the coming weather probably will be. For the public this foreknowledge is as yet only a completely frozen asset. However, it is immediately made available by every practicable speedy method.

Then soon, in a few hours at most, the whole process must be gone over again, for the weather keeps right on changing all the time, and that, too, in many cases, without much notice. What's the sense, our philosopher asks, in all this bother to know what the weather probably will be a few hours hence, or even the next day? What's coming is coming and all the king's horses and all the king's men can change it not a whit.

Well, even the most stolid (or stubborn) philosopher now and then wants to know whether or not he should carry an umbrella, while many who are not philosophers, or, at any rate, not that sort, want to know about the coming weather for many good reasons besides the natural desire to keep comfortably dry. If, for instance, one has an engagement to play a round of golf with a friend, the weather permitting, he certainly wants to know about that "permission "—whether the morrow will be fair or foul. So, too, does the baseball fan, and everyone else who indulges in any sort of an outdoor sport, whether as active participant or noisy rooter.

To one on a fishing trip the state of the morrow's weather is a matter of much concern. He of course wants a good catch, one to brag about, and fish are choosy about the weather, hardly biting at all when it is raining. Curious, but probably not because they are afraid then of getting wet, as the wag explained it.

Similarly, a fox club that exults in a lively chase is anxious about the weather to expect; they want the air to be cool and damp, because when the forest and field are dry the fox leaves no scent behind him for a hound dog to follow, and the sport is all off. And this is a matter of such grave importance that even rainmakers have been employed to save the situation!

Can we have tomorrow's dedication, celebration, or whatever it may be, out under the trees where it would be most enjoyable, or must it be indoors? Is an anxious question often asked by many a church and college, asked of the official weather service since it alone has the abundant information necessary to a fairly reliable answer to this question, and knows how to make the most of it. True, no far-flung weather service would be established and maintained just for such occasional needs as these; but, being established, it can be, and is, so used to an aggregate very great benefit of all the public, a benefit the public would not now forego for a good deal more than such a service actually costs for these and its many far greater benefits of every and sundry sort all combined.

The holiday, or weekend, tripper, desirous of a safe and a pleasant journey, does well to get all the foreknowledge of the coming weather he can before starting, or even deciding whether or not to go. To him, and his name is legion, this help alone is worth all the whole service costs. Whether he is encouraged to go and have a good time, or stay at home and play safe, the information is worth a great deal to him.

And there are several special warnings of great local value, in addition to the general forecasts for relatively large areas. In forested regions, for example, especially those having valuable stands of timber, it is very necessary to know when special precautions should be taken to prevent the start and spread of fires.

When the forest duff is dry, a forecast of low humidity is a call for precaution, and if at the same time there is to be wind strong enough even to be called a breeze, the demand is for great care. No fires of any kind, sawmill, kitchen, cigarette, should be lighted in the region so affected, and all the available fire fighters should be on the *qui vive* for an alarm anywhere at any moment.

But even when furnace fires are all drawn, meals served cold and cigarettes put away, there still may, and often does, remain a serious fire hazard from the nature of the weather itself of which it is necessary to have the fullest possible foreknowledge. One must know whether to expect local thunderstorms, and if so, where and when, for many a forest fire is started by lightning, especially by the lightning of "dry" thunderstorms, that is, thunderstorms in which most, and often all, the rain is evaporated before it reaches the ground. Many fires are started by lightning under these conditions and some of them are exceedingly destructive.

It is not the business, of course, of a weather service either to locate forest fires or to fight them when and where they occur, but it is the business of every such service to keep the forest rangers or guardians and protectors of our valuable woodlands, promptly and fully informed as to impending conditions every time there is an approach of fire-hazard weather — low humidity, high temperature, considerable wind and probability of thunderstorms.

It also is the business of a weather service to issue warnings of the very antithesis of forest fires, that is, of floods. A few of these, but relatively very few, are owing to breaks in retaining walls, incident, some of them, to earthquakes.

The others, and that is nearly all, have their origin in heavy rain or the rapid melting of snow, or, occasionally, both together. Of course once a flood has started it does not require a weather service to predict that it is going on down the river, but just how a particular flood is going to behave is best judged from a careful study of what previous similar floods have done in the same river.

This study logically could be made, and commonly is made, by the weather service that covers the region affected. In order that his warnings may be substantially accurate and therefore of great benefit the flood forecaster must know the state of the soil, because when either saturated or frozen, most of any additional rain that falls is promptly drained off over the surface instead of soaked up, as it largely is when the ground is dry.

He must know also whether, and if so to what depth, the area involved is covered with snow, and what is on for the morrow's temperature and rainfall—facts, all of them, best known to the weather service, which service therefore is the logical one to be charged with the important work of issuing flood warnings.

Another class of local weather warnings, local in application, but important to all who eat fruit for breakfast and vegetables for dinner, are those called for by the orchardist and the trucker, who, to have any, must protect their products from frost at every critical period; and at all stages, bud to basket, from scale and scald, rust and rot, worm and weevil, and endless other pests whose preying leadeth not to prayers.

When frost or freeze is imminent the orange grove, for instance, must be kept warm by many small open fires, a troublesome and costly business. Woe betide, therefore, the watchman who cries "wolf, wolf," when there is no wolf; and woe betide him, too, when the danger is real and he warns not at all or warns too late. But, as just stated, frost is not the only hazard to a wholesome breakfast and a bountiful dinner. Every growing and spoiling pest also must be kept in check, and that implies the labour and expense of abundant spraying or dusting with whatever the appropriate substance may be.

Obviously, however, all this labour is lost when it is immediately followed by a rain that washes clean again every leaf that had been so carefully treated. Clearly, too, here is where the weather man can and does render valuable service, by his timely warning that rain is but a few hours off, or, equally, by the assurance that for a day or two, at least, nothing will happen that to any extent will undo the work in hand.

Orchardists and trucksters, however, are not the only cultivators of the soil who on occasions find it highly profitable to consult the weather man. The harvester long ago learned to heed the farmer's admonition to "make hay while the sun shines." He knows this is the right thing to do, but is uncertain when to mow, when the weather will remain fair long enough for the hay to cure and be stacked or put under cover.

This valuable information he can get, and if prudent does get, from the weather service, for the mere asking, or, rather, by merely listening at the right time to the weather broadcast. Not only in hay harvesting is the farmer thus helped by a knowledge of the coming weather, but in every other sort of harvesting as well, even to the cutting of timber, and in practically everything else he does, for his, above all others, is a weather-controlled industry.

Obviously, as just explained, the growing and successful harvesting of a crop, especially of fruit or vegetables, are greatly at the mercy of the weather, but once that crop is safely stored, so far as any effect on it is concerned the weather, it would seem, might go whistle. But not so in the case of onions, apples or anything else of a like perishable nature.

A sudden cold wave may catch a car load, or many car loads, of such produce on its way to market and spoil every bit of it. As a matter of fact it seldom happens that a large shipment of a perishable commodity is so caught, owing chiefly to timing the date of shipment according to the ap- proaching weather. But this is not all the benefit we get from a warning in time that a cold wave is near and coming fast.

That is when prudent folks drain all their inadequately protected water pipes; do what they can to prevent freezing in their automobile radiators; get in more coal, if the supply in the bin is low; spread extra blankets on their beds; get out heavy clothing and take all the other precautions they naturally would take if headed for the Arctic for, just as bad, the Arctic is headed for them.

Clearly, then, cold-wave warnings are of great benefit to everyone in the path of the on-rushing frigid flood, and that is why weather services in the countries subject, as is the United States, to sudden outbursts of vast masses of frigid air, issue special warnings of their coming. They can not be stopped, but we can guard against their harmful effects. To be fore-warned should be and is, for the wise, to be fore-armed.

A wholly different class of weather forecasts from the above, but equally valuable, pertains to the warning of coming storms on lakes and along ocean coasts. Indeed it was the havoc wrought by such storms that gave the first adequate incentive to the establishment of a national weather service in Europe, and that, similarly, led to the founding of a like service in the United States.

On November 14, 1854, during the Crimean war, a hurricane on the Black Sea caused to the French-English allied fleet serious damage, most, if not all, of which easily could have been avoided if only the coming of the storm had been known to the captains a few hours before it broke. Through correspondence shortly after the time of this disaster the storm that caused it was followed hour by hour across Europe, and the obvious inference drawn that with the telegraphic facilities then available almost any sort of a weather service could have warned the vessels of the coming storm in ample time for them to have gotten into safe harbors.

The good this wind did was to blow into the consciousness of many European officials the realization that a national weather service was possible, and might be worth its cost. Such a service soon was begun in France, and a little later essentially duplicated in other countries.

In the United States it was repeated damage to vessels on the Great Lakes that finally set the American weather ball rolling. Its first official, or congressional, push was made on February 9, 1870, and this was it:

Be it resolved by the Senate and House of Representatives of the United States of America in Congress assembled, That the Secretary of War be, and he hereby is, authorized and required to provide for taking meteorological observations at the military stations in the interior of the continent and at other points in the States and Territories of the United States, and for giving notice on the northern lakes and on the seacoast, by magnetic telegraph and marine signals, of the approach and force of storms.

Mighty was, and still is, the growth of that ball. It gathered vast material from all the oceans and the ends of the earth, and then through kites, balloons, aeroplanes and the robot radiosonde invaded the blue heavens for information about the state and condition of the atmosphere up there.

A foreknowledge of coming storms on inland seas and along all coasts has long been recognized, then, as a matter of great importance; more important, indeed, than would be the like knowledge for places far from land. This is because many more vessels sail lakes and coast shores than travel in mid-ocean; because coast-trade vessels commonly are smaller, frailer and less able to withstand the buffeting of high winds and huge waves than are the ocean-going steamers; and because reefs and rocks, the major hazards to any ship, chiefly are along the shore line and not far out at sea.

Hence, on the receipt of warnings of coast storms, small vessels in harbor remain there until the danger is over, while those outside hasten to shelter as speedily as possible. In both cases lives are saved and property protected—ample compensation, and more, for the total cost of the organized service that issues the warning information.

While it is true, as stated, that out at sea most atmospheric disturbances are comparatively harmless, there nevertheless is one that frequently is dangerous wherever encountered, dangerous in open water, perilous along the coast. This is that violent, rotating tropical storm of numerous aliases; hurricane in the West Indies, cyclone on the Indian Ocean, typhoon on the China Sea, baguio in the Philippines, and many another as determined by place and language.

But wherever it occurs and whatever called, all who get warning of its coming do everything they can to protect themselves from its fury; especially do coastal dwellers seek high ground to avoid being swept away by the invariably accompanying and all-engulfing tidal waves.

It will be a relief now to turn for a while from dangers to delights, from ocean storms to one of the ocean's most pleasing products, snow in the mountains. Keeping a close record of the amount of solid water stored up in great mountain reservoirs is one of the duties of every weather service that covers regions in which this phenomenon occurs to a marked extent.

It is from these data that fairly reliable and highly valuable estimates are made of the amount of electric power and quantity of irrigation that can be had during the coming summer through run-off from the region surveyed, values that vary rather widely from year to year. Evidently this foreknowledge must be, as it actually is, highly valuable to the growers of whatever crops, and the consumers of electric power for whatever purpose.

Many of the valleys and slopes in and adjacent to these great snowy mountains afford our best plots for certain experiments in, and the study of, agricultural meteorology, that is, the relation of the growth and maturing of useful plants to kind of weather and time of its occurrence.

This evidently is an exceedingly difficult problem since no sort of weather will stay put, even the wind bloweth where it listeth, and the other elements are just as fickle, save alone in regions where nothing grows without irrigation and nearly everything flourishes with it. Here sunshine is abundant and water can be supplied, much or little, at any desired hour and chosen stage of growth. But the partially controlled conditions in an irrigated region do not duplicate Nature elsewhere, and it is mostly elsewhere that crops are grown.

Hence, that we may know as much as possible about agricultural meteorology, it is necessary that continuous and accurate records be kept of the weather in the fields where the crops under investigation are grown. The problems in this study are so complex and the data necessary so irregular that only general conclusions can be drawn from them.

But even so they are valuable, since from them it often is easy to infer what the crop probably will be long before the crop is gathered. This, to be sure, may aid the grain speculator, but we can't help that. The rain, we are told, falls alike on the just and the unjust.

Through many years, as is well known, all meteorology, observational, theoretical and applied, though covering the whole earth, was confined to but the length and the breadth thereof. Finally, however, it ambitiously took in the remaining, or vertical, dimension, went aloft with, and to serve, the aeroplane. A little indeed had already been done in this third dimension with kites and balloons, but it was only a feeble prelude to the work of today in the free air.

The aviator's needs are for definite information and reliable forecasts. He has little interest in, and less need for, long-range forecasts even if it were, which it is not, feasible to make them; but those that look only two or three hours ahead he does greatly need, and wants no maybes about them in any respect, neither as to place, when nor what. "Probably showers tonight or tomorrow in the western portion of the state," might do for most people, but not for the aviator.

He insists on knowing whether or not there will be a thunderstorm, for instance, over a particular landing field today at 3: 30 P.M. the exact time he is due to arrive there. He wants to know, too, what the speed and direction of the wind will be at various levels along the course he is about to take; how high the clouds will be; how far ahead he will be able to see; where there will be fog, if any.

Some of the information he needs must be so up-to-the-minute that he has to get it in mid-flight by radio. The gathering and interpreting of mid-air weather data for whatever purpose, civil, as an aid in the forecasting of surface weather, or military, commonly is called aeronautical meteorology, because its most extensive use is by the aviator to whom it is of vital importance.

Just as those parts of meteorology, those states, conditions and actions of the atmosphere, that are of greatest concern to the aviator, are collectively, together with the practical work therewith, called aeronautical meteorology, so, for a like reason, those of special use in military operations, and their practical applications thereto, are called military meteorology.

Thus in war, a foreknowledge of the coming weather of the next day or two, and longer if possible, is of great importance, since an attacking army is more likely to succeed when the ground is dry than when it is so wet as to bog down the cavalry —horse or tank. Indeed, so great is the influence of the weather on army movements that January and February long have been known as Russia's major generals.

Similarly, April and May might well be called Finland's chief defence strategists owing to the way they bog and mire the country beyond passage. And everywhere wars are much more summer diversions than winter sports, as someone remarked facetiously, but with correct weather implication.

The occurrence of a heavy fog may create confusion or even change the whole course of a battle, and therefore when probable is taken into account by a wise commander.

Foreknowledge of the strength and direction of the wind is essential in the use of poison gases. Obviously, because a strong wind would render them harmless by dissipation and dilution, while just the right sort of a drift, if shifted to the wrong direction after the gas was set free, as such winds sometimes do shift, would become a disastrous boomerang.

And there are other occasions to have full knowledge of the winds. Thus, in static warfare especially, it often becomes desirable to silence a disturbing enemy battery that can not be seen. The first thing to do towards that end, obviously, is to find out exactly where it is, since shooting blindly is just wasting aminunition. The basic data for this purpose are the exact instants the same sound made by a cannon at the hidden battery reaches certain known stations.

The rest is quick manipulation of ingenious devices based on mathematical calculations, plus the proper correction for the drift of the sound incident to the direction and speed of the wind through which it came. This correction is vital since accurate shooting at the wrong thing is worse than blind shooting at the right thing. To determine this correction it is necessary to know all about the movements and temperature of the air between the source of the sound and the places where its arrivals are detected.

These facts the meteorologist must supply. And even yet, an

accurate gun (corrected for the rotation of the earth at the local latitude) aimed directly at the target commonly would miss the mark, owing to drift of the projectile by the winds through which it passed. From a knowledge of these winds, level to level, which information it is the business of the army meteorologist at all times to have, the final correction in the pointing of our own gun can be made, and the enemy's disturbing battery promptly silenced—maybe.

The ultimate basis of meteorology, however, is a thing far more pleasing to contemplate than wars or even rumors of war, that is, solar radiation, without which the surface of the earth would be far colder than liquid air, and every sort of life of which we have any knowledge, both animal and vegetable, wholly impossible thereon.

Owing to the many and vital effects of this radiation as a whole, and of its every portion from the utmost ultra-violet to the deepest infra-red, it deserves, and is receiving, searching measurement with various sorts of delicate apparatus, and profound studies, both theoretical and experimental.

Still another special meteorological problem of much interest and great difficulty is that of evaporation. At every evaporating surface there appears also to be condensation. Commonly, however, only the difference between the two is measured and the net result called evaporation where the loss is greater than the gain, and condensation where the gain is greater than the loss.

The rate of total, not net, evaporation of water from a given area, or, in other words, the total, not net, loss of water per minute from a given unit area, such as one square inch, of surface varies with the temperature of the water; with the kind and amount of impurity dissolved in it; with the extent of the water surface beyond that under consideration; with the shape of this surface; with the humidity of the space near it; with the strength of the wind across it; and of course, with the intensity of any radiation that may be incident upon it.

The extent to which any one of these factors alone would affect the rate of evaporation is fairly well known, but in Nature they all operate at the same time and so irregularly that it commonly is impossible to compute from them satisfactory values of evaporation. It therefore nearly always is just measured as a sum-total effect with but little regard to the various separate controlling factors involved. This suffices to determine, at least roughly, the loss of water by evaporation from given lakes and reservoirs.

Evaporation from plants and from soil also are matters of importance, but here the conditions are still more complicated and variable than are those that affect evaporation from a free water surface, so complicated indeed and variable that it seems quite useless to try to compute the rate of evaporation from them. Here direct measurements, and they alone, give useful values.

There is yet one other branch of meteorology that needs to be specifically and separately mentioned, medical meteorology. It is not concerned with the cause or causes of any weather phenomenon, nor with any phase or kind of weather prediction, but essentially with correlations between kinds of weather and the prevalence of this or that malady. It noted the fact that hay fever is essentially an autumn affliction long before its cause was known; that pulmonary troubles are more prevalent in the spring of the year than in midsummer; that cold damp weather is especially favorable to rheumatism; and many other such associations.

It tells us that the great Greek physician, Hippocrates, was right when he told us, 2350 years ago, that fresh air and sunshine are the best of medicines. It shows that health resorts commonly are healthful, but does not explain why. It divides a country into separate regions on the basis of the kinds of diseases therein and their prevalence —or rarity, if local pride so insists. It is much more medical climatology, which name it also has, than medical meteorology, but whatever it is called, it is a useful compendium for physician and patient too.

In addition to all the foregoing, there are innumerable other things, ranging from a to z and back again, that are more or less affected by the weather. Not only must hay, especially alfalfa and red clover, be made while the sun shines, but many other things, such as peaches, prunes and raisins are at least seriously injured if rained on while in the drying, or curing, process.

Many manufactured products too, from cement to chocolate candy, are made at the mercy of the weather. Even a blast furnace is affected by the condition of the air, requiring less fuel and getting hotter when the humidity is low than when it is high.

Advertising also is largely weather determined, overshoes and umbrellas when rain is predicted for the morrow; skates on the eve of a cold wave, and so on through a long list of things appropriate to rain or shine, snow or freeze. Finally, a business transaction is quickest consummated when the wind is in the Northwest.

Chapter 6

Glaciers and Sea Ice

Fifteen thousand years ago, much of what is now continental shelf was dry land. Sea level was 130 meters lower than it is today. Where was the 45 million cubic kilometers of water (about 12 billion billion gallons!) this difference in sea level represents?

It was locked up in the great glaciers of the ice age. In the warmer period in which we find ourselves today, the water that remains frozen as ice still far exceeds all other reservoirs of fresh water on Earth. Were this ice to melt, sea level would rise another 80 meters.

It is important in this connection to distinguish between *glaciers* and *sea ice*. A glacier is a sheet of ice formed *on land* when accumulated snow is compressed and hardened into ice by the weight of overlying layers. Sea ice is ice formed from seawater; it floats on the ocean's surface, although it may be attached to land at its edges.

The glaciers that usually come to mind when one hears this term are the scattered "mountain and valley glaciers" that occur throughout the world's high mountain ranges — the Himalayas, Andes, Rockies, and Alps, for example. The larger glaciers of this variety are some tens of kilometers long, a kilometre or more across, and a few hundred meters thick.

These glaciers are constantly in motion, being fed by snowfall at their surfaces in the higher elevations and moving downhill as the deep layers, under great pressure, flow as a plastic solid. (Perhaps the easiest way to visualize what is going on in such flow, which as we noted earlier also occurs in rock in Earth's mantle, is to consider the plastic solid to be an extremely viscous fluid.)

The speed of advance varies along the length of the glacier, but is typically 100 meters per year or more in the main body of the larger mountain glaciers. The advance of such glaciers is terminated by melting of the tongue of the glacier at the lower end.

By far the greatest part of the world's inventory of ice — more than 99 per cent — is tied up in a second kind of glacier, the land ice or ice sheets, that cover the bulk of the Greenland and Antarctic landmasses. The formation of such a sheet requires an arctic climate, sufficient precipitation, and fairly flat land.

The ice layer that results covers essentially the entire landscape in a gently sloping dome, interrupted only by a few projecting mountain peaks. The Greenland ice sheet covers an area of 1.74 million square kilometers (about 80 per cent of the total area of Greenland) and has an average thickness of about 1600 meters (5250 ft). The Antarctic ice sheet covers 13 million square kilometers with ice up to 4000 meters thick (13,000 ft) and averaging perhaps 2300 meters. About 91 per cent of the world's ice thus is in the Antarctic sheet and about 9 per cent in the Greenland sheet. T

hese ice sheets, like mountain and valley glaciers, are in motion, carrying to the sea the ice formed from precipitation in the central regions. Typical speeds are some tens of meters per year on the ice sheet proper, but they can be much higher — hundreds, and even thousands, of meters per year — where certain glacial tongues meet the sea.

Where the ice sheets meet the sea in broad expanses, they may extend into the ocean as more-or-less floating ice shelves, from tens to hundreds of meters thick. In the Antarctic, these shelves reach widths of hundreds of kilometers. The largest, the Ross Ice Shelf, covers more than 500,000 square kilometers. Icebergs originate when great masses of ice break off from the tips of glacial tongues or the edges of ice shelves and are carried away (often into shipping lanes) by currents.

Sea ice, as distinguished from floating extensions or pieces of glaciers, is formed by the freezing of seawater on the ocean surface. The North Pole ice pack, with a mean extent of about 10 million square kilometers, is a collection of slabs of sea ice floating on the Arctic Ocean. In winter, these slabs are frozen together and attached to land at various points around the ocean's periphery.

In summer, some of the slabs break apart and are separated by narrow strips of open water, and the southern limit of the ice retreats northward. The sea ice, which begins to form at − 2°C, is porous, and the enclosed cavities often contain water saltier than seawater. Glacial ice, by contrast, consists of fresh water, being simply compacted and recrystallized snow.

The maximum thickness of sea ice is only between 3 and 5 meters. Once it reaches this thickness, the layer of ice insulates the underlying water so well that no more can freeze — heat is supplied from the deeper water faster than the surface layer can lose it through the ice. (Ice is a poor conductor of heat and snow an even poorer one, which is why snow igloos stay so warm inside.)

If the average thickness of the North Pole ice pack is 2 meters, it contains less than one-hundredth as much ice as the Greenland ice sheet. Of course, melting of the sea ice would have no direct effect on sea level, even if the volume of this ice were much greater; the ice is floating, thus displacing an amount of water equal to its weight, so it is already contributing exactly as much to the level of the oceans as it would if it melted. Aside from the sea-level issue, however, which relates solely to glacial ice sheets, the sea ice has great importance for climate.

Fig. Glacier

The Hydrologic Cycle

Although oceans and ice caps contain some 99.3 per cent of all the water on Earth the fraction of 1 per cent residing at any given time in the atmosphere, in lakes and streams, and in soil and subsurface layers plays unique and important roles. The flow of water on the surface is a major determinant of the configuration of the physical environment. Soil moisture is essential to most terrestrial plant life.

The stocks and flows of ground and surface water are major links in the transport and cycling of chemical nutrients and important determinants of what kinds and intensities of human activity can be supported in what locations. And water in the atmosphere has several functions that are central to shaping climates.

The set of processes that maintain the flow of water through the terrestrial and atmospheric branches of the hydrosphere is called the *hydrologic cycle*. The cycle includes all three physical states of water — liquid, solid (ice and snow), and gas (water vapour). It also includes all of the possible transformations among these states — *vaporization*, or evaporation (liquid to gas); *condensation* (gas to liquid), *freezing* (liquid to solid); melting, or *fusion* (solid to liquid); and *sublimation* (gas to solid, or the reverse).

The principal flows in the hydrologic cycle are:

- *Evaporation* of water from the surface of the oceans and other bodies of water, and from the soil;
- *Transpiration* of water by plants, the result of which is the same as that of evaporation — namely, the addition of water vapour to the atmosphere;
- Horizontal *transport* of atmospheric water from one place to another, either as vapour or as the liquid water droplets and ice crystals in clouds;
- *Precipitation,* in which atmospheric water vapour condenses (and perhaps freezes) or sublimates and falls on the oceans and the continents as rain, sleet, hail, or snow;
- *Runoff,* in which water that has fallen on the continents as precipitation finds its way, flowing on and under the surface, back to the oceans.

Because it is difficult and not particularly useful to distinguish between the contributions of evaporation and transpiration on the continents, these two terms are often lumped together as *evapotranspiration.*

The magnitudes of these flows, averaged over all the continents and oceans and expressed in thousands of cubic kilometers of water per year. These magnitudes are based on the assumption that the various components of the hydrosphere are in equilibrium, which is at least a good first approximation.

That is, on a year-round average, inflows and outflows for the atmosphere, the oceans, and the continents all balance. (For example, in thousands of cubic kilometers, the atmosphere receives 62 + 456 = 518 as evaporation from the surface and gives up 108 + 410 = 518 as precipitation.)

The magnitude of the flows in the hydrologic cycle is more readily grasped if one thinks of the flows in terms of the equivalent depth of water, averaged over the surface area involved.

In these terms, the world's oceans annually lose to evaporation a layer of water 1.26 meters deep (about 4 feet) over their entire surfaces, gaining back 1.14 meters from precipitation and 0.12 meters from the discharge of rivers and groundwater.

The continents receive precipitation each year equivalent to a layer of water 0.73 meters (29 in) deep over their entire surface areas, of which 0.42 meters is lost to evaporation and 0.31 meters makes up the runoff.

Combining the foregoing information on equilibrium flows with the information on stocks in the hydrosphere permits us to estimate the average residence time of water in the different parts of the cycle. These residence times, which are of great importance in analyzing the transport of pollutants, as well as nutrients, by the hydrologic cycle.

There is an enormous range, from the average nine days a water molecule spends in the atmosphere between being evaporated from the surface and falling again as precipitation, to the 10,000 years a molecule of water typically spends as ice between falling in a snowflake on the Antarctic ice sheet and rejoining the ocean with the melting of an iceberg.

It is also important to remember that there are large deviations from the average in any given category — a water molecule may fall in a raindrop not nine days but an hour after being evaporated from Earth's surface; another may wander not two weeks but two years in the delta of the Amazon River before reaching the sea.

Nevertheless, the average residence times can provide useful insights into a variety of important problems, and the approach can be refined whenever information more pertinent than global averages is available.

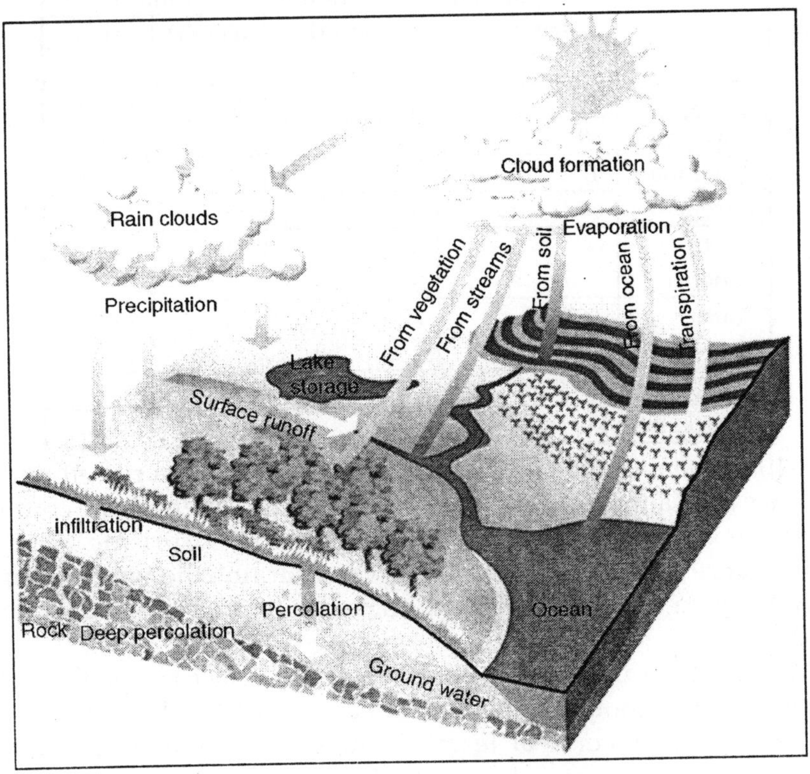

Fig. The Hydrologic Cycle.

Table. Residence Times of Water Molecules in the Hydrologic Cycle

Location	Residence time
Atmosphere	9 days
Rivers (typical speed, 1 m/sec)	2 weeks
Soil moisture	2 weeks to 1 year
Largest lakes	10 years
Shallow groundwater (speed, 1-10 m/day)	10s to 100s of years
Mixed layer of oceans (1st 150 m)	120 years
World ocean	3000 years
Deep groundwater	up to 10,000 years
Antarctic ice cap	10,000 years

The size of the runoff (the difference between precipitation and evapotranspiration) is a measure of how much water is potentially available for domestic and industrial uses by society (including dilution and removal of wastes) and for the other functions that flowing water performs.

Table. Average Water Balance of the Continents

	Precipitation (cm/yr)	Evaporation (cm/yr)	Runoff	
			(cm/yr)	(km /yr)
Africa	69	43	26	7,700
Asia	60	31	29	13,000
Australia	47	42	5	380
Europe	64	39	25	2,200
North America	66	32	34	8,100
South America	163	70	93	16,600
Includes entire USSR.				
Source: Budyko, M. I., p. 227.				

2-10 the remarkable fact that South America has a runoff per unit of surface area almost three times that of North America, the continent with the next greatest runoff. It is perhaps not so surprising, then, that the discharge of the Amazon River, which drains the wettest third of South America, amounts to about a seventh of the runoff of the entire world.

Much of the runoff on the continents takes place not on the surface but beneath it. Although the quantities can only be estimated, it is clear that most rivers receive at least as much of their flow through the ground as from flow over the ground; and a certain amount of water reaches the oceans via flowing aquifers at the edges of the continents without ever joining a surface river at all.

Water beneath the land's surface is called *soil moisture,* or soil water, when it is distributed in the first meter or so of soil (a zone defined by the depth of penetration of the roots of most plants.) Below the zone of soil moisture is an intermediate zone where the water percolates downward through open pores in the soil and rock; and below this is the *water table,* marking the surface of the body of *groundwater* that saturates the soil or rock in which it finds itself, filling all pores and spaces in the soil or rock completely.

The groundwater extends downward until it is limited by an impermeable layer of rock. In some circumstances, there are successive layers of groundwater (*aquifers*) separated by impermeable layers of rock. The absolute lowest limit of groundwater is probably about 16 kilometers from the surface, where the pressure is so great that all pores are closed and any rock becomes impermeable.

Most groundwater is flowing, albeit very slowly [10 meters per day (m/d) in coarse gravel near the surface, more commonly 1 m/d, and much more slowly at greater depths.] At 1 meter per day, of course, it takes almost three years to move 1 kilometre.

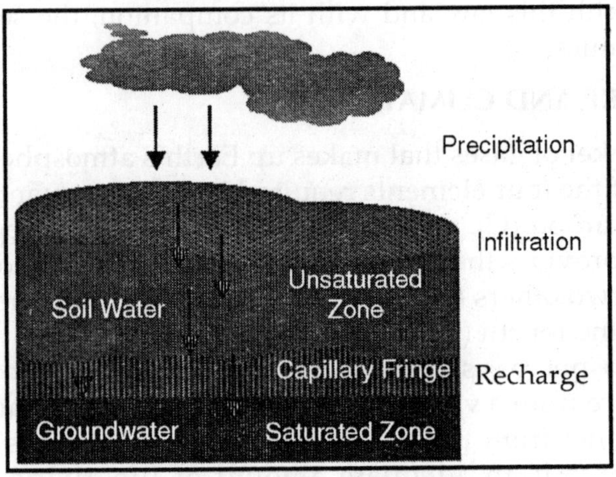

Fig. Zones of Subsurface Water.

The energy that drives the hydrologic cycle is energy from the sun — indeed, this function is the largest single user of the solar energy reaching Earth's surface. The reason so much energy is required is that it takes a great deal of energy to evaporate water 2250 joules per gram at the boiling point of 100°C and 2440 joules per gram at Earth's average surface temperature of 15°C.

It takes fifty times as much energy to evaporate a gram of water as it does to lift it to an altitude of 5 kilometers. The energy used to evaporate the water is stored as latent heat of vaporization which is released to the environment as heat whenever and wherever the water vapour condenses into liquid.

Thus, energy delivered by the sun at one point on Earth's surface may be released high in the atmosphere over a point 1000 kilometers away. This mechanism of redistributing energy by the transport and condensation of water vapour is a major determinant of Earth's climate.

As noted above, the energy the sun supplies at the time of evaporation reappears as heat at the time of condensation.

Similarly, the smaller amount of solar energy that does the work of lifting the water vapour against the force of gravity appears as frictional heat when falling droplets of condensed vapour collide with molecules of air and when rushing mountain streams rub against their rocky beds.

That all the energy the sun supplies to terrestrial processes comes back again in one form or another is not coincidence or quirk, but an illustration of the first law of thermodynamics — the law of conservation of energy. Further excursions into the machinery of the physical world — and of human technology — will require some familiarity with this law and with its companion, the second law of thermodynamics.

ATMOSPHERE AND CLIMATE

The blanket of gases that makes up Earth's atmosphere has many functions. Of the four elements required in greatest quantity by living organisms [carbon (C), oxygen (O), hydrogen (H), nitrogen (N)] the atmosphere provides the main reservoir of one (N), the most accessible reservoir of two others (C, O), and an essential link in the continuous recycling of the fourth (H, in the form of H_2O).

The atmosphere is substantial enough to protect the organisms on Earth's surface from a variety of harmful particles and radiations that reach the planet from the sun and from space, but it is transparent enough to permit an adequate amount of life-giving sunlight to penetrate to that surface.

Acting as a thermal insulator, the atmosphere keeps Earth's surface much warmer, on the average, than it would be if there were no atmosphere. And the stirrings of the atmosphere, transporting energy and moisture from one place to another, are a major part of the patterns of climate so important to the character and distribution of life.

For simplicity, we begin our investigation of the atmosphere by ignoring its internal vertical and horizontal motions and considering its properties as a static body of gas.

Air

The term *air* refers to the particular mixture of gaseous compounds making up the atmosphere. The average composition of this mixture, not including water. An important property of gases is that a given number of molecules at a given temperature and pressure will occupy almost exactly the same volume, regardless of the mass or size of the molecules. This property has led to the use of the somewhat confusing terms *per cent by volume or fraction by volume* to describe the relative abundance of the various constituents of gaseous mixtures.

That is, if three-quarters of the molecules in a container of fixed volume are gas A and one-quarter are gas B, one can think of gas A as "occupying" three-fourths of the volume and gas B as "occupying" one-fourth. (What is happening in reality, of course, is that both gases, mixed together, occupy the whole volume, with gas A accounting for three-fourths of the pressure in the volume and gas B accounting for one-fourth of the pressure.)

We will use the term *molecular fraction* (number of molecules of a constituent divided by the total number of molecules in the mixture), because it is unambiguous and works for solids and liquids as well as gases.

Table. Average Composition of Clean Dry Air

Constituent	Symbol	Molecular weight	Molecular fraction of air	Mass fraction of air
Nitrogen	N_2	28	0.7809	0.755
Oxygen	O_2	32	0.2095	0.232
Argon	Ar	40	0.0093	0.013
Carbon Dioxide	CO_2	44	320 ppm	486 ppm
Neon	Ne	20	18 ppm	12 ppm
Helium	He	4	5.2 ppm	0.7 ppm
Methane	CH_4	16	2.9 ppm	1.6 ppm
Krypton	Kr	84	1.1 ppm	3.2 ppm
Nitrous Oxide	N_2O	44	0.5 ppm	0.8 ppm
Hydrogen	H_2	2	0.5 ppm	0.03 ppm
Ozone	O_3	48	0.01 ppm	0.02 ppm

Many processes in nature and in technology involve the transformation of energy from one form into others. For example, light from the sun is transformed, upon striking a meadow, into thermal energy in the warmed soil, rocks, and plants; into latent heat of vaporization as water evaporates from the soil and through the surface of the plants; and into chemical energy captured in the plants by photosynthesis.

Some of the thermal energy, in turn, is transformed into infrared electromagnetic radiation heading skyward. The imposing science of thermodynamics is just the set of principles governing the bookkeeping by which one keeps track of energy as it moves through such transformations. A grasp of these principles of bookkeeping is essential to an understanding of many problems in environmental sciences and energy technology.

The essence of the accounting is embodied in two concepts known as the first and second laws of thermodynamics. No exception to either one has ever been observed. The first law, also known as the law of conservation of energy, says that energy can neither be created nor destroyed. If energy in one form or one place disappears, the same amount must show up in another form or another place.

In other words, although transformations can alter the *distribution* of amounts of energy among its different forms, the *total* amount of energy, when all forms are taken into account, remains the same. The term *energy consumption,* therefore, is a misnomer; energy is used, but it is not really consumed.

One can speak of fuel consumption, because fuel, as such, does get used up. But when we burn gasoline, the amounts of energy that appear as mechanical energy, thermal energy, electromagnetic radiation, and other forms are exactly equal all together to the amount of chemical potential energy that disappears. The accounts must always balance; apparent exceptions have invariably turned out to stem from measurement errors or from overlooking categories. The immediate relevance of the first law for human affairs is often stated succinctly as, "You can't get something for nothing."

If the amount of stored work never diminishes, how can we become worse off? One obvious answer is that we can become worse off if energy flows to places where we can no longer get at it — for example, infrared radiation escaping from Earth into space. Then the stored work is no longer accessible to us, although it still exists.

A far more fundamental point, however, is that *different kinds of stored work are not equally convertible into useful, applied work.* We can therefore become worse off if energy is transformed from a more convertible form to a less convertible one, even though no energy is destroyed and even if the energy has not moved to an inaccessible place. The degree of convertibility of energy — stored work — into applied work is often called *availability.*

Energy in forms having high availability (that is, in which a relatively large fraction of the stored work can be converted into applied work) is often called high-grade energy.

Correspondingly, energy of which only a small fraction can be converted to applied work is called low-grade energy, and energy that moves from the former category to the latter is said to have been degraded.

Electricity and the chemical energy stored in gasoline are examples of high-grade energy; the infrared radiation from a light bulb and the thermal energy in an automobile exhaust are corresponding examples of lower-grade energy. The quantitative measure of the availability of thermal energy is temperature.

More specifically, the larger the *temperature difference* between a substance and its environment, the more convertible into applied work is the thermal energy the substance contains; in other words, the greater the temperature difference, the greater the availability. A small pan of water boiling at 100°C in surroundings that are at 20°C represents considerable available energy because of the temperature difference; the water in a swimming pool at the same 20°C temperature as the surroundings contains far more total thermal energy than the water in the pan, but the availability of the thermal energy in the swimming pool is zero, because there is no temperature difference between it and its surroundings.

With this background, one can state succinctly the subtle and overwhelmingly important message of the second law of thermodynamics: *all physical processes, natural and technological, proceed in such a way that the availability of the energy involved decreases.* (Idealized processes can be constructed theoretically in which the availability of the energy involved stays constant, rather than decreasing, but in all real processes there is *some* decrease.

The second law says that an *increase* is not possible, even in an ideal process.) As with the first law, apparent violations of the second law often stem from leaving something out of the accounting.

In many processes, for example, the availability of energy in some *part* of the affected system increases, but the decrease of availability elsewhere in the system is always large enough to result in a net decrease in availability of energy overall.

What is consumed when we use energy, then, is not energy itself but its availability for doing useful work.The statement of the second law given above is deceptively simple; whole books have been written about equivalent formulations of the law and about its implications. Among the most important of these formulations and implications are the following:

- In any transformation of energy, some of the energy is degraded.
- No process is possible whose sole result is the conversion of a given quantity of heat (thermal energy) into an equal amount of useful work.
- No process is possible whose sole result is the flow of heat from a colder body to a hotter one.
- The availability of a given quantity of energy can only be used once; that is, the property of convertibility into useful work cannot be "recycled."
- In spontaneous processes, concentrations (of anything) tend to disperse, structure tends to disappear, order becomes disorder.

That Statements 1 through 4 are equivalent to or follow from our original formulation is readily verified. To see that statement 5 is related to the other statements, however, requires establishing a formal connection between order and availability of energy.

This connection has been established in thermodynamics through the concept of *entropy,* a well defined measure of disorder that can be shown to be a measure of unavailability of energy, as well. A statement of the second law that contains or is equivalent to all the others is: *all physical processes proceed in such a way that the entropy of the universe increases.*

Consider some everyday examples of various aspects of the second law. If a partitioned container is filled with hot water on one side and cold water on the other and is left to itself, the hot water cools and the cold water warms — heat flows from hotter to colder. Note that the opposite process does not violate the first law, conservation of energy. That it does not occur illustrates the second law.

Indeed, many processes can be imagined that satisfy the first law but violate the second and therefore are not expected to occur. As another example, consider adding a drop of dye to a glass of water. Intuition and the second law dictate that the dye will spread, eventually coloring all the water concentrations disperse, order (the dye/no dye ar- rangement) disappears. The opposite process, the spontaneous concentration of dispersed dye, is consistent with conservation of energy but not with the second law.

A more complicated situation is that of the refrigerator, a device that certainly causes heat to flow from cold objects (the contents of the refrigerator — say, beer — which are made colder) to a hot one (the room, which the refrigerator makes warmer).

But this heat flow is not the *sole* result of the operation of the refrigerator: energy must be supplied to the refrigeration cycle from an external source, and this energy is converted to heat and discharged to the room, along with the heat removed from the interior of the refrigerator. Overall, availability of energy has decreased, and entropy has increased.

One illustration of the power of the laws of thermodynamics is that in many situations they can be used to predict the maximum efficiency that could be achieved by a perfect machine, without specifying any details of the machine! (Efficiency may be defined, in this situation, as the ratio of useful work to total energy flow.)

Thus, one can specify, for example, what *minimum* amount of energy is necessary to separate salt from seawater, to separate metals from their ores, and to separate pollutants from auto exhaust without knowing any details about future inventions that might be devised for these purposes.

Similarly, if one is told the temperature of a source of thermal energy — say, the hot rock deep in Earth's crust — one can calculate rather easily the maximum efficiency with which this thermal energy can be converted to applied work, regardless of the cleverness of future inventors. In other words, *there are some fixed limits to technological innovation, placed there by fundamental laws of nature.*

More generally, the laws of thermodynamics explain why we need a continual input of energy to maintain ourselves, why we must eat much more than a pound of food in order to gain a pound of weight, and why the total energy flow through plants will always be much greater than that through plant-eaters, which in turn will always be much greater than that through flesheaters.

They also make it clear that *all* the energy used on the face of the Earth, whether of solar or nuclear origin, will ultimately be degraded to heat. Here the laws catch us both coming and going, for they put limits on the efficiency with which we can manipulate this heat. Hence, they pose the danger that human society may make this planet uncomfortably warm with degraded energy long before it runs out of high-grade energy to consume.

Occasionally it is suggested erroneously that the process of biological evolution represents a violation of the second law of thermodynamics. After all, the development of complicated living organisms from primordial chemical precursors, and the growing structure and complexity of the biosphere over the eons, do appear to be the sort of spontaneous increases in order excluded by the second law.

The catch is that Earth is not an isolated system; the process of evolution has been powered by the sun, and the decrease in entropy on Earth represented by the growing structure of the biosphere is more than counterbalanced by the increase in the entropy of the sun.

It is often asked whether a revolutionary development in physics, such as Einstein's theory of relativity, might not open the way to circumvention of the laws of thermodynamics. Perhaps it would be imprudent to declare that in no distant corner of the universe or hither to unexplored compartment of subatomic matter will any exception ever turn up, even though our intrepid astrophysicists and particle physicists have not yet found a single one.

But to wait for the laws of thermodynamics to be overturned as descriptions of everyday experiences on this planet is, literally, to wait for the day when beer refrigerates itself in hot weather and squashed cats on the freeway spontaneously reassemble themselves and trot away.

A *mole* of any substance is 6.02 × 10 molecules, and the mass of a mole is equal to the molecular weight of the substance in grams. For example, the mass of a mole of nitrogen gas (N_2) is 28 grams. It is often convenient to speak of air as if it were a single substance; the term a *mole of air* means 6.02 × 10 molecules, of which 78.09 per cent are nitrogen molecules, 20.95 per cent are oxygen molecules, and so on. Such a collection of molecules has a mass of about 29 grams, which is called the *molecular weight of air*. (These definitions will be of importance later in interpreting what is meant by pollution standards expressed in different ways.)

Although nitrogen and oxygen comprise 99 per cent of dry air, the trace constituents carbon dioxide (CO_2) and ozone (O_3) play exceedingly important roles because of the special properties of these molecules. Methane, nitrous oxide, and hydrogen also have roles in atmospheric chemistry and physics, albeit smaller ones.

Argon, helium, krypton, and neon, by contrast, are chemically inert, monatomic gases, whose presence in the atmosphere is of interest only as resources for certain applications in technology.

Water Vapour

The water content of the atmosphere varies greatly from place to place and time to time. Three commonly used measures of water content are *absolute humidity, specific humidity,* and *relative humidity.* Absolute humidity is the mass of water vapour per unit volume of air, and it varies from almost zero over the driest deserts to around 25 grams per cubic meter over jungles and tropical seas.

Specific humidity is the mass of water vapour per unit mass of air. (A closely related term, the *mixing ratio,* is the mass of water vapour mixed with each unit mass of *dry* air.) Relative humidity, usually expressed as a percentage, is the ratio of the actual molecular fraction of water vapour in air to the molecular fraction corresponding to saturation at the prevailing temperature.

(*Saturation* refers to the condition that ensues if air is left for a long time in a sealed container partly filled with pure water; the number of molecules of water vapour per unit volume of air under these circumstances depends only on the temperature.) Relative humidity usually is between 0 and 100 per cent, but under special circumstances (supersaturation) it can significantly exceed 100 per cent.

Under ordinary circumstances, water vapour in the atmosphere begins to condense into droplets of liquid, forming clouds, as soon as the relative humidity exceeds 100 per cent by even a small amount. The process of condensation is greatly facilitated by the virtually universal presence in the atmosphere of small particles that provide surfaces where the condensation can commence.

Called *condensation nuclei* when they perform this function, these particles include salt crystals formed by the evaporation of sea spray, dust raised by the wind, ash from volcanoes and forest fires, decomposed organic matter, and, of course, particles produced by various technological activities.

Even in "unpolluted" air, particles that might serve as condensation nuclei are seemingly abundant in absolute terms (more than 100 particles per cubic centimeter), but the extent of condensation and precipitation apparently are related to specific physical characteristics of the condensation nuclei as well as to their number.

The molecular fraction of water vapour corresponding to saturation increases as temperature increases — warm air can "hold" more water vapour than cool air. Accordingly, there are two ways in which relative humidity can be raised from less than 100 per cent to more, initiating condensation and perhaps precipitation:

- More water vapour can be added to the air by evaporation from an exposed water surface.
- The air can be cooled so that the vapour content corresponding to saturation falls.

At a given vapour content (a fixed specific humidity), the temperature at which the relative humidity reaches 100 per cent is called the *dew point*. Addition of water vapour to the air by evaporation is a slow process, but cooling of the air can be very rapid. Rapid cooling to below the dew point is the mechanism immediately responsible for most condensation phenomena — the appearance of dew and fog at night as air is cooled by radiation of heat to the night sky; formation of clouds and rain in updrafts as the air is cooled by expansion; and formation of beads of water on the outside of a pitcher of ice water on a hot day, as air adjacent to the pitcher is cooled by contact with the cold surface.

PRESSURE, TEMPERATURE, AND VERTICAL STRUCTURE

The pressure exerted by the atmosphere on objects at Earth's surface is essentially equal to the weight of the overlying air, which at sea level amounts on the average to 10.3 metric tons per square meter (14.7 lb/in). This amount of pressure, defined as I atmosphere, is the same as would be exerted at sea level by a column of water about 10 meters high (33 ft) or a column of mercury 760 millimeters (mm) high (29.92 in).

This means that the mass of the atmosphere is only equivalent to that of a 10-meter layer of water covering Earth and that pressure under water increases by the equivalent of I atmosphere for every 10 meters of depth. The usual metric unit for the measurement of atmospheric pressure is the millibar; 1 millibar is 100 newtons (N) per square meter the definition of the newton and I atmosphere is 1013.25 millibars.

Atmospheric pressure is not ordinarily perceived as a force because it acts equally in all directions; organisms are not crushed by it because the gases and liquids in tissue are also at atmospheric pressure, so the inward and outward forces balance. Pressure becomes perceptible as a (painful) pressure *difference* if the pressure outside an organism changes more rapidly than the interior pressure can accommodate (an example would be the pain in one's ears associated with a rapid change in altitude). Unlike water, whose density at the bottom of the deepest ocean trenches at pressures of hundreds of atmospheres is only a few percentage points higher than its density at the surface, the air in the atmosphere is highly compressible — that is, density increases markedly as pressure increases.

Indeed, air behaves very much like a "perfect gas," for which pressure (p), density (ρ), and temperature (T) are related by the equation $p = \rho RT$ For this equation to be valid, the temperature T must be measured with respect to *absolute zero*, the temperature at which there is no molecular motion. Temperature measured from this zero point, which is the same for all substances, is called *absolute temperature*, and the corresponding unit of measurement in the metric system is the degree kelvin. The R in the equation is the gas constant, which for dry air equals 287 joules per kilogram per degree kelvin. According to the perfect gas equation, the density of dry air varies in direct proportion to pressure, if temperature is held constant. Because the atmosphere is compressible, its mass is concentrated in the lower layers. Forty per cent of the air in the atmosphere lies below the altitude of the summit of Mount Whitney in California's Sierra Nevada range (4.4 km) and two-thirds lies below the altitude of the summit of Mount Everest (8.9 km). The density of air at an altitude of 12 kilometers, where most subsonic jet airliners fly, is about one-fifth the density at sea level. The average variation of pressure and temperature with altitude above sea level.

The atmosphere is subdivided into horizontal layers according to the pattern of temperature variation. The lowest layer, called the troposphere, is characterized by a rather uniform average rate of temperature decline with altitude of 6.4°C per kilometre. Almost all the atmospheric phenomena that govern climate take place in the troposphere. The top of the troposphere is called the *tropopause*, where the temperature decline stops and a layer of uniform temperature at about -55°C commences. The tropopause is typically found at an altitude of from 10 to 12 kilometers, but it ranges from a low of 5 or 6 kilometers at the poles to around 18 kilometers at the equator.

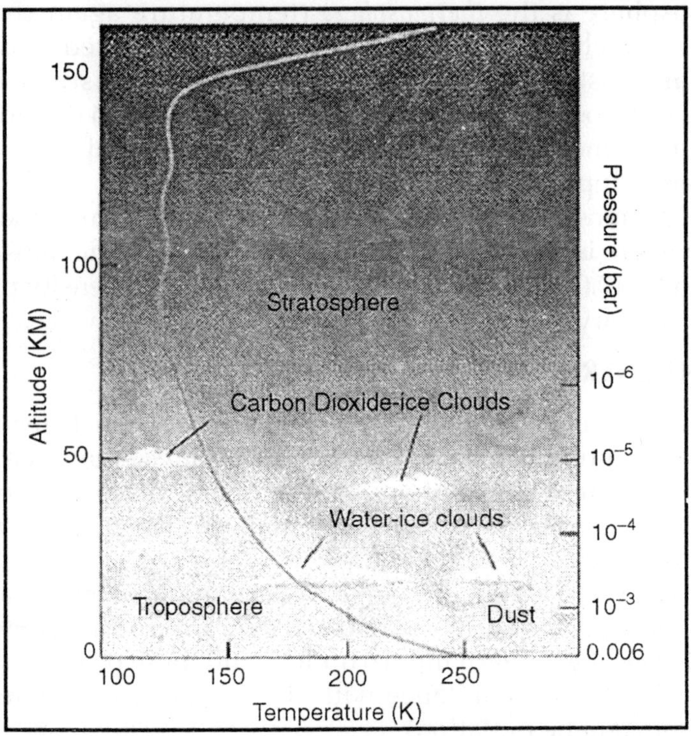

Fig. Variation of Atmospheric Temperature and Pressure.

The *stratosphere* extends from the tropopause up to the stratopause (about 50 km) and is characterized over much of this interval by temperatures increasing with altitude (reaching almost 0°C at the stratopause). The gaseous composition of the stratosphere is essentially the same as that at sea level, with two significant exceptions.

First, there is very little water vapour in the stratosphere; the mixing ratio is typically two or three parts per million (ppm), or 1000 to 10,000 times less than is common near sea level. Second, there is a great deal more ozone in the stratosphere than in the troposphere; the maximum molecular fraction of ozone is 10 ppm near 25 km altitude, or 1000 times more than the average for the whole atmosphere.

The air pressure and density at the top of the stratopause are on the order of a thousandth of the values at sea level. Above the stratosphere lies the *mesosphere* (to about 90 km), wherein the temperature again decreases with altitude. The composition of the mesosphere remain much like that of the lower layers, except for certain trace constituents such as water vapour and ozone.

The troposphere, stratosphere, and mesosphere together are called the *homosphere*, referring to their relatively uniform composition. Above the mesosphere is the *thermosphere* (temperature again rising with altitude), which contains the *heterosphere* (so named because the molecular constituents are there separated into distinct layers of differing composition) and the *ionosphere* (referring to the presence of free electrons and the positively charged ions from which the electrons have been stripped).

The thermosphere has no well defined upper limit; its density at 100 kilometers is around one-millionth of atmospheric density at sea level, and by 10,000 kilometers it has faded off to the density prevailing in interplanetary space.

RADIANT ENERGY FLOW IN THE ATMOSPHERE

What accounts for the complicated way in which temperature changes with altitude in the atmosphere? The answer involves the way in which different atmospheric constituents interact with radiant energy arriving from the sun and with radiant energy trying to escape Earth into space.

The same processes, of course, determine how much and what kinds of energy reach Earth's surface, so they are crucial in determining the conditions that govern life. To understand these processes requires at least a modest acquaintance with the character of radiant energy (or electromagnetic radiation).

The energy in the electromagnetic radiation reaching the top of

Earth's atmosphere from the sun is distributed over a range of wavelengths. One can determine from such a graph that about 9 per cent of the total incoming energy is in the ultraviolet part of the electromagnetic spectrum, 41 per cent is in the visible part of the spectrum, and 50 per cent is in the infrared part.

A significant part of this energy is prevented from reaching Earth's surface by gaseous constituents of the atmosphere that are opaque to certain wavelengths. This opacity is due not to reflection but to absorption (the energy in the radiation is absorbed by the gas molecules, warming the atmospheric layers where these processes take place).

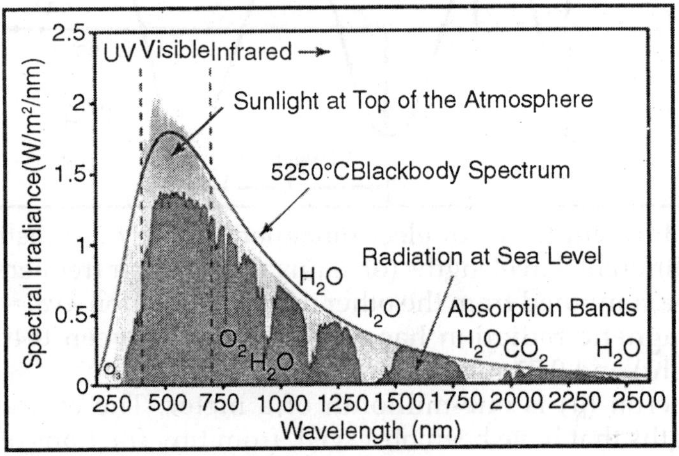

Fig. Solar Irradiance Spectrum

The depletion of incoming radiation by absorption in atmospheric gases. The main results are that ultraviolet solar radiation with wavelengths less than 0.3 microns (μ) is almost completely absorbed high in the atmosphere, and infrared solar radiation is substantially depleted through absorp-Light, X-rays, radio waves, infrared radiation, and radar waves are all variations of the same thing phenomena with the interchangeable names *electromagnetic radiation, electromagnetic waves*, and *radiant energy*.

Energy in this form travels at the speed of light and the words *in vacuum* imply requires no material medium to support the energy flow. This property contrasts with the other, much slower forms of energy transport (conduction and convection) which do require a medium. Convection involves the bulk motion of matter; conduction involves molecular motion. What moves in the case of electromagnetic radiation is a combination of electric and magnetic fields of force.

For many purposes, it is useful to visualize this pattern as a traveling wave. There the curve denotes the spatial pattern of the strength of the electric or magnetic field, and the pattern is moving to the right with speed *c*. (Actually, there should be separate curves for the electric and magnetic fields, but this detail need not trouble us here.)

The *wavelength* (λ) is the distance between successive crests or troughs. At any fixed point along the path of the wave, the field is seen to oscillate with *frequency* (v) related to wavelength and the speed of light by the relation $v = c/\lambda$.

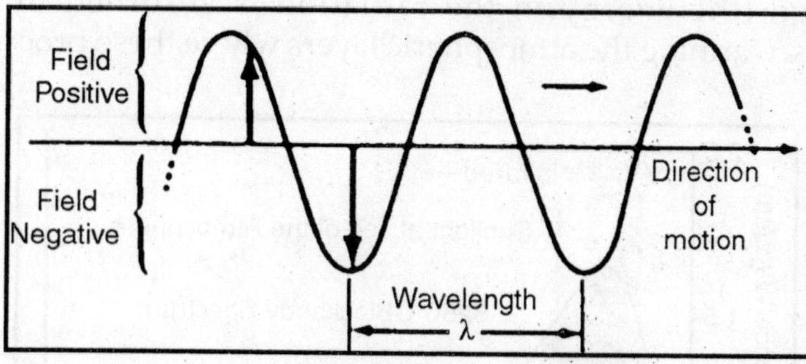

The different forms of electromagnetic energy are distinguished by their different wavelengths (or, equivalently, their frequencies, since one can be computed from the other using the relation $\lambda v8, = c$). Visible electromagnetic radiation has wavelengths between 0.40 microns (violet light) and 0.71 microns (red light).

A micron (μ) is one-millionth of a meter. The entire range of wavelengths that have been observed, from tiny fractions of a micron to tens of kilometers, is called the *electromagnetic spectrum*. Some types of electromagnetic radiation occupying different parts of the spectrum.

Type of Radiation	Wavelength Range
radio	1-10 m
radar (microwaves)	1-30 cm
infrared	0.71-100 μ
visible	0.40-0.71 μ
ultraviolet	0.10-0.40 μ
X-rays	10 -10 μ

The way electromagnetic radiation interacts with matter depends in a complicated manner on the wavelength of the radiation, on its intensity (energy flow per unit of area, measured — say — in watts per square meter and on the properties of the matter. Radiant energy that encounters matter may be transmitted, reflected, or absorbed.

To the extent that radiation of a given wavelength is transmitted, the material is said to be transparent to that wavelength; to the extent that the radiation is reflected or absorbed, the material is said to be opaque to that wavelength. Most materials are transparent to some wavelengths and opaque to others. Many gases, for example, are rather opaque to most ultraviolet wavelengths but transparent to radio waves, visible light, and X-rays. Human flesh is opaque to visible light but transparent to X-rays.

Often, transmission, reflection, and absorption all take place at once. Consider a glass window in the morning sunlight: the glint off the window indicates that reflection is taking place; the room behind the window is illuminated and warmed, so there is certainly transmission; and the window itself gets warm, so absorption is happening, too.

Both transmission and reflection can be *direct* or *diffuse*. Direct means that a beam of electromagnetic radiation arriving from a single specific direction is sent on or sent back in a single specific direction. Diffuse means that the incoming beam is split up and sent on or sent back in many different directions.

The phenomenon that produces diffuse transmission and reflection is called *scattering*. All light reaching the ground on a completely overcast day is diffuse light that has been scattered by the water droplets in the cloud layer.

All matter that is warmer than absolute zero can *emit* radiation. (The "machinery" by which emission occurs has to do with the behaviour of the electrons in matter. We will not dwell on this machinery here, summarizing instead only the main results of its operation.) Some substances emit only radiation of certain wavelengths, and all substances can absorb only the wavelengths they can emit. A body that absorbs all the radiation that hits it in all wavelengths is called a *blackbody*, a useful idealization that is approached in the real world but never quite reached.

A blackbody is not only a perfect absorber, but also a perfect emitter; the amount of electromagnetic energy emitted at any given wavelength by a blackbody of a specified temperature is the theoretical maximum that any real body of that temperature can emit at that wavelength. The total amount of radiation emitted by a blackbody is proportional to the fourth power of the body's absolute temperature. The wavelength at which a blackbody emits most intensely decreases in inverse proportion to the absolute temperature — that is, the higher the temperature, the shorter the wavelength.

The characteristics of the sun as a source of electromagnetic radiation are very closely approximated by those of a blackbody with a temperature of 5800°K. The wavelength at which its emission is most intense is 0.5 microns, corresponding to blue-green light.

Table. Absorption of Solar Radiation by Atmospheric Gases

Wavelength Range (μ)	*Fate of Radiation*
UNTRAVIOLET	
Less than 0.12	All absorbed by O_2 and N_2 above 100 km
0.12-0.18	All absorbed by O_2 above 50 km
0.18-0.30	All absorbed by O_3 between 25 and 50 km
0.30-0.34	Part absorbed by O_3
0.34-0.40	Transmitted to Earth almost undiminished
VISIBLE	
0.40-0.71	Transmitted to Earth almost undiminished
INFRARED	
0.71-3	Absorbed by CO_2 and H_2O, mostly below 10 km

The atmosphere's gases are almost completely transparent to the visible wavelengths, where the intensity of solar radiation reaches its peak and to the "near-ultraviolet" radiation, with wavelengths just shorter than the visible. Significantly, the only atmospheric gas that is opaque to ultraviolet radiation between 0.18 and 0.30 microns in wavelength is ozone.

Without the trace of ozone that exists in the stratosphere, this radiation would reach Earth's surface where it could be extremely disruptive to the life forms that evolved in the absence of these wavelengths. The absorption of ultraviolet radiation by ozone in the stratosphere has a second important effect — it produces the stratospheric heating that causes temperature to increase with altitude in this layer of the atmosphere.

Some of the incoming radiation that is not absorbed by atmospheric gases is scattered by them. Some of the scattered radiation returns to space and some reaches the ground as diffuse solar radiation. The physics of scattering by the molecules in air is such that blue light is scattered much more than is red light.

The sky appears blue because what reaches the eye is mostly scattered light and hence — owing to the preferential scattering of blue wavelengths by air molecules — mostly blue. That the sun appears red at sunset is precisely the same phenomenon at work; then one is looking at direct-beam radiation — the unscattered part -from which the blue has been removed (by scattering) and in which mainly the red remains.

Absorption and scattering of solar radiation are done by *aerosols* as well as by gases. An aerosol is a suspension of solid or liquid particles in a gas. Fog, clouds, smoke, dust, volcanic ash, and suspended sea salts are all aerosols. Whether a given aerosol acts mainly as an absorber or mainly as a scatterer depends on the size and composition of the particles, on their altitude, and on the relative humidity of the air they are in. Certainly, the aerosols that interact most strongly with solar radiation are clouds, which at any given time cover about half Earth's surface with a highly reflective layer.

The reflectivity of a surface or a substance is called its *albedo* (formally, albedo equals reflected energy divided by incoming energy). This property depends not only on the characteristics of the surface, but also on the angle at which the incoming radiation strikes the surface (the angle of incidence). The albedo is much higher at shallow angles of incidence than at steep (nearly perpendicular) angles. The albedo of clouds ranges from 0.25 (thin clouds, perpendicular incidence) to more than 0.90.

Averaged around the year and around the globe, the amount of energy that penetrates to Earth's surface is about half of what strikes the top of the atmosphere. Of that which penetrates, somewhat more than half is diffuse and somewhat less than half is direct. Upon reaching the surface, this energy meets several fates.

Some is reflected, with the albedo varying from land to water and from one type of vegetation to another — and, of course, depending on the angle of incidence. Some is absorbed by the melting or sublimation of ice or snow, or the vaporization of water. Some is absorbed to warm the surface and objects on or under it. And a tiny fraction is captured and transformed into chemical energy by the process of photosynthesis in plants.

Table. Albedo of Various Surfaces

Surface	Albedo
Snow	0.50-0.90
Water	0.03-0.80
Sand	0.20-0.30
Grass	0.20-0.25
Soil	0.15-0.25
Forest	0.05-0.25

Note that the average albedo of the Earth-atmosphere system is about 0.28, of which two-thirds is accounted for by clouds. Now from the first law of thermodynamics, which says energy can neither be created nor destroyed, and from elementary considerations of stocks and flows, one can draw some important conclusions about energy flow in the Earth-atmosphere system.

First, the rate at which Earth's atmosphere and surface are absorbing solar energy must be matched by the rate at which the system loses energy, or else the amount of energy stored on the surface and in the atmosphere would be steadily changing. In other words, if the system is to be in equilibrium, outflow must equal inflow. Lack of equilibrium, if it occurred, would mean a changing stock of energy in the Earth-atmosphere system.

This could manifest itself as an upward or downward trend in mean temperature (a changing stock of thermal energy), or in mean absolute humidity or mean volume of ice and snow (a changing stock of latent energy of vaporization and fusion), or in mean quantity of organic matter (a changing stock of chemical energy). Of course, different parts of Earth's surface and atmosphere at different times of year are out of equilibrium — temperature, humidity, snow cover, and quantity of vegetation generally change dramatically with seasons.

Inflow and outflow of energy at a given time and place in general do not balance. But they *must* balance on a year-round average for the whole globe or else there will be year-to-year changes in the *mean* values of these indices of stored energy.

Such changes have occurred in the past as cooling periods leading into ice ages and warming periods leading out of them. These changes and the potential for further ones, with or without human influence. As a first approximation, however, it is reasonable to assume that inflows and outflows of energy are in balance on the global, annual average.

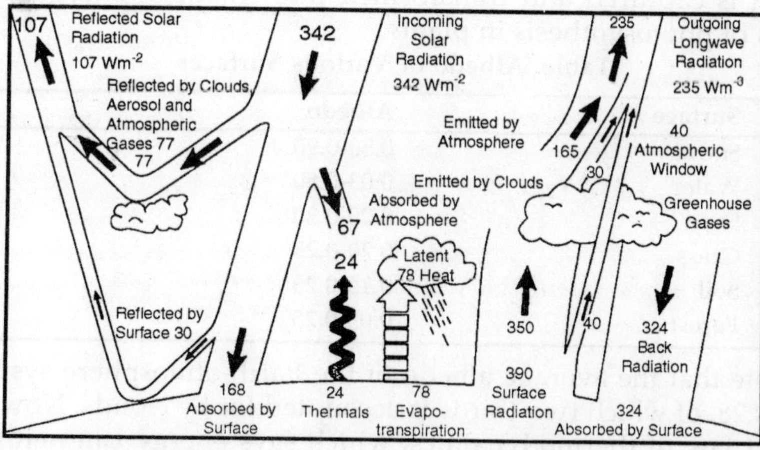

Fig. The Fate of Incoming Solar Radiation.

The second conclusion is that the same considerations of inflow, outflow, and equilibrium that apply to the Earth-atmosphere system as a whole must apply separately to its components. If the atmosphere is to be in equilibrium in a global, time-averaged sense, it must be losing energy at the same rate it is receiving it. The same must hold for Earth's surface. A net accumulation of energy in the atmosphere (gas molecules, clouds, dust) equal to 22 per cent of the solar energy that strikes the top of the atmosphere and a net accumulation of energy at the surface equal to 47 per cent of total incoming solar energy.

What is the fate of this absorbed energy, amounting in all to 69 per cent of the solar input? The answer is this: after running the machinery of winds, waves, ocean currents, the hydrologic cycle, and photosynthesis, the energy is sent back to space as *terrestrial radiation.*

Terrestrial radiation refers to the electromagnetic radiation emitted by Earth's surface and atmosphere in accordance with the principles: the amount of energy radiated per unit of area is proportional to the fourth power of the absolute temperature of the radiating substance, and the wavelength of most intense radiation is inversely proportional to the absolute temperature. If the Earth-atmosphere system were a blackbody at temperature, T, the exact relation for the rate of emission of radiation, S, would be:

$$S = \sigma T^{\cdot}$$

If S is measured in watts per square meter and T is measured in degrees kelvin, then the proportionality constant a (the Stefan-Boltzmann constant) has the numerical value $5.67 \times 10^{-\cdot}$ On the assumption that the energy returned to space as terrestrial radiation should exactly balance the 69 per cent of incoming solar energy that is absorbed, one can compute the effective blackbody temperature of the Earth-atmosphere system to be about 255°K (–18°C or about 0°F). This is the temperature a perfect radiator would have to have in order to radiate away into space the same amount of energy per unit of area as the Earth-atmosphere system actually does radiate away, on the average.

Actually, the Earth-atmosphere system closely resembles a blackbody radiator, so the above temperature is not unrealistic. Why, then, is that temperature so much lower than the observed mean surface temperature of 288°K (15°C, or 59°F)?

The reason is that most of the terrestrial radiation actually escaping from the Earth atmosphere system is emitted by the atmosphere, not the surface. Most of the terrestrial radiation emitted by the warmer surface does not escape directly to space because the atmosphere is largely opaque to radiation of these wavelengths.

The wavelength of radiated energy, increases as the temperature of the radiator decreases: Earth, being much cooler than the sun, emits its radiation at longer wavelengths. The peak intensity of terrestrial radiation occurs at a wavelength of around 10 microns, which is in the infrared part of the spectrum, in contrast to solar radiation's peak intensity at a wavelength of about 0.5 microns in the visible part of the spectrum.

The opaqueness of the atmosphere to outgoing infrared radiation from the surface is due mainly to three atmospheric constituents: carbon dioxide, water vapour, and clouds. Carbon dioxide absorbs infrared radiation in a narrow band of wavelengths around 3 microns, in another narrow band near 4 microns, and in the wavelengths between 12 and 18 microns.

Water vapour absorbs infrared radiation in narrow bands around 1, 1.5, and 2 microns, and in broader ones from 2.5 to 3.5 microns, from 5 to 8 microns, and from 15 microns through the remainder of the infrared.

Clouds are very much like blackbodies in the entire infrared part of the electromagnetic spectrum — they absorb most of the infrared radiation that reaches them. The infrared radiation absorbed by carbon dioxide, water vapour, and clouds is subsequently *reradiated* by these substances — much of it being sent back in the direction of Earth's surface and some escaping into space.

The atmosphere, therefore, through the properties of clouds, water vapour, and carbon dioxide, acts as a thermal blanket that keeps Earth's surface about 33°C warmer than it would be without these constituents.

Because of this effect of clouds, all else being equal, clear nights are colder than cloudy nights; in the absence of clouds, more infrared radiation leaving the surface escapes directly to space without being intercepted. The additional thermal-blanket effect of water vapour and carbon dioxide is sometimes called the *greenhouse effect*.

This is because glass, like carbon dioxide and water vapour, is relatively transparent to visible radiation but more opaque to infrared radiation. Light enters a greenhouse more readily than heat can escape, a situation resembling that in the atmosphere. The *main* reason a greenhouse is warmer inside than outside, however, is that the glass prevents convection from carrying away sensible heat and latent heat of vaporization. Thus, the term *greenhouse effect* is not entirely appropriate for the role of carbon dioxide and water vapour in the atmosphere.

The flows of terrestrial radiation are expressed as a percentage of solar energy striking the top of the atmosphere and computed as a global annual average. Note that the rate at which infrared radiation is emitted from Earth's surface actually exceeds the rate of solar input at the top of the atmosphere and is far larger than the rate at which solar energy is directly received at the surface.

This large radiation output and the high temperature responsible for it are made possible by the large flow of infrared radiation reradiated to the surface by the atmosphere, as discussed earlier. Any alteration of the composition of the atmosphere by pollution can have an important influence on this downward reradiated energy flow.

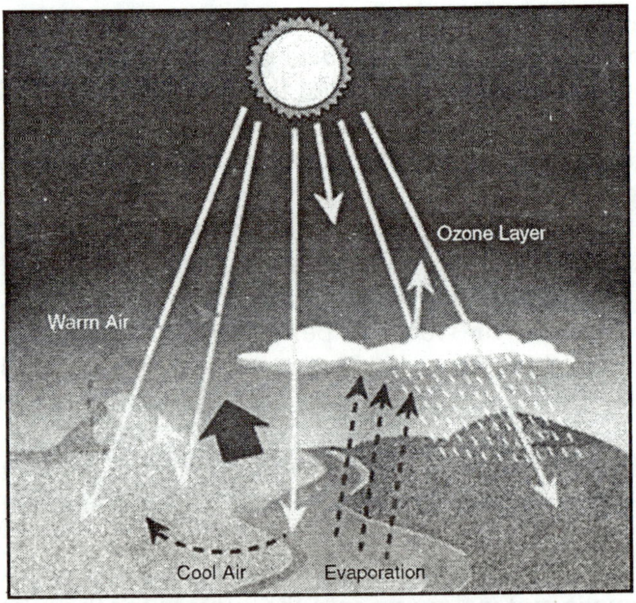

Fig. Flows of Terrestrial Radiation.

Two important energy flows other than radiation are represented by broken lines. The larger of these flows is the transfer of energy from the surface to the atmosphere as latent heat of vaporization. That is, energy that has been used to evaporate water at the surface moves into the atmosphere in the form of the latent heat of vaporization that is associated with the water vapour. This energy is eventually surrendered to the atmosphere as sensible heat when the water vapour condenses.

The magnitude of this surfaceto-atmosphere energy flow is equal to almost a quarter of the solar energy flow reaching the top of the atmosphere. The second major non-radiative flow (perhaps a fifth as large as that of latent heat) is the transfer of sensible heat from the surface to the atmosphere by conduction — that is, the warming of the air by contact with the surface. Sensible heat transported across the interface of surface and air in this way then moves upward in the atmosphere by convection.

ATMOSPHERIC ENERGY BALANCE AND VERTICAL MOTIONS

With the information in the preceding section, one can construct average energy balances for Earth's surface, for the atmosphere, and for the Earth-atmosphere system as a whole. Such a set of balances both in terms of the percentage of solar energy flow at the top of the atmosphere and in watts per square meter of Earth's surface area. Energy flow per unit of area is called flux.

The solar flux through a surface perpendicular to the sun's rays at the top of the atmosphere is called the solar constant and is equal to about 1360 watts per square meter [1.95 calories (cal)/cm /minute (min), or 1.95 langleys/min, in units often used by meteorologists]. As a sphere (to a very good approximation), Earth's total surface area is 4 times the area of the cross section it presents to the sun's rays.

Hence, the average amount of solar energy reaching the top of the atmosphere per square meter of Earth's surface is just one-fourth of the solar constant, or 340 watts per square meter. The average solar flux reaching Earth's surface is 173 watts per square meter of horizontal surface. Solar flux measured this way at the surface is often termed *insolation*.

The difference between the amount of energy incident on Earth's surface as radiation (solar and terrestrial) and the amount leaving as radiation is called by meteorologists the *net radiation balance*, or sometimes just net radiation. This amount of energy flow averages 105 watts per square meter flowing into the surface for the whole globe, is of special meteorological significance because it is the amount of energy available for the climatically crucial processes of evaporation of water and surface-to-air transfer of sensible heat. Many discussions of human impact on climate use the net radiation balance, rather than the incident solar energy, as the yardstick against which civilization's disturbances are measured.

The pattern of Earth-atmosphere energy flows that has been described here provides the explanation for the observed temperature distribution in the lower atmosphere. Basically, the atmosphere is heated from the bottom — by direct contact with warm land or water surfaces, by the release of latent heat of vaporization when water vapour condenses (virtually entirely in the troposphere and mostly in its lowest third), and by the absorption of terrestrial infrared radiation in water vapour, clouds, and carbon dioxide.

Water vapour and clouds are most abundant in the lower troposphere, so that is where a large part of the absorption takes place. These effects, differentially heating the atmosphere near the bottom, produce the general trend of decreasing temperature with altitude in the troposphere.

Chapter 7

Nutrient Cycles

The main flow of energy that helps shape conditions on Earth's surface comes from space; and when the energy's work here is done, to space it returns. With respect to energy, then, Earth is an open system. With respect to its chemical endowment, however, Earth is a closed system.

That is, the amounts of carbon, hydrogen, oxygen, iron, gold, and other elements in the planet-atmosphere system do not change with time; the chemical arrangement and physical distribution of these elements can and do vary, but essentially nothing enters and nothing leaves the system.

The elements in this closed system that are essential to life are called *nutrients*. They can be divided into three categories: the 4 main chemical building blocks of living matter (carbon, oxygen, hydrogen, nitrogen); 7 *macronutrients*, of which smaller but still significant quantities are required for life; and 13 *micronutrients*, or trace elements, of which tiny quantities perform essential functions.

The suitability of any terrestrial or aquatic environment for the support of life depends on the availability of nutrients in appropriate forms and quantities. The processes that govern this availability (or lack of it) are known collectively as *nutrient cycles*, because of the way the individual basic stocks of physical material move cyclically through the living and nonliving parts of the physical world.

These cycles function in support of life not merely by making nutrients continuously available — in other words, by maintaining the fertility of the environment -but also by limiting the accumulation of material in quantities, forms, and places in which it would damage organisms.

It is important to recognize in this connection that the same elements and compounds that serve as nutrients for some organisms in some concentrations are often toxic for other organisms, or even for the same organisms at higher concentrations. For example, molecular oxygen (O_2) is toxic to anaerobic organisms and, at high enough concentrations, even to mammals. Ammonia (NH_3) is an important source of nutrient nitrogen for many plants but is toxic to people. Hydrogen sulfide (H_2S) is a nutrient for certain types of bacteria but is extremely toxic to mammals.

Table. The Chemical Elements of Life

	Symbol	Name	Atomic number	Atomic weight
Basic building blocks				
	H	Hydrogen	1	1
	C	Carbon	6	12
	N	Nitrogen	7	14
	O	Oxygen	8	16
Macronutrients				
	Na	Sodium	11	23
	Mg	Magnesium	12	24
	P	Phosphorus	15	31
	S	Sulfur	16	32
	Cl	Chlorine	17	35
	K	Potassium	19	39
	Ca	Calcium	20	40
Trace elements				
	F	Fluorine	9	19
	Si	Silicon	14	28
	V	Vanadium	23	51
	Cr	Chromium	24	52
	Mn	Manganese	25	55
	Fe	Iron	26	56
	Co	Cobalt	27	59
	Cu	Copper	29	64
	Zn	Zinc	30	65
	Se	Selenium	34	79
	Mo	Molybdenum	42	96
	Sn	Tin	50	119
	I	Iodine	53	127

DYNAMICS OF NUTRIENT CYCLING

The steps that make up all nutrient cycles involve two basic processes: physical transport and chemical transformation. They are:

- The hydrologic cycle, including evaporation and precipitation as well as the flow of streams, groundwater, and ice.
- Winds.
- Ocean currents.

- Geologic movement, especially the upward and downward motions at the boundaries of tectonic plates and geologic uplifting on continents.

A fifth agent of physical transport that is important in a few circumstances is the movement of organisms. For example, fish-eating birds that deposit their excrement on land provide a significant pathway by which nitrogen and phosphorus are transferred from the sea to the land. Fish such as salmon, that feed mostly in the ocean but migrate up freshwater rivers to spawn and die, perform a similar function, as do the ocean-caught fish consumed by continent-dwelling human beings.

How the various mechanisms of transport link the principal "compartments" of the physical world. Obviously, the ease (and, hence, the rate) with which a given substance can move through these pathways depends on its physical state (solid, liquid, gas), on other physical properties (solubility, volatility), and on the possibility of chemical transformations that can change those properties.

Such chemical transformations take place in living organisms and in the nonliving environment. Each of the compartments indicated often is divided into subcompartments for the purpose of tracing nutrient flows. Thus, for example, it is customary to consider three subcompartments of material in both the soil and the oceans: living organic matter, dead organic matter, and the inorganic medium itself.

These subcompartments can be further subdivided — for example, organic matter, into plants, animals, and microorganisms; and seawater, into the well-mixed surface layer and the stratified deeper layers.

Finally, within each subdivision of each sub-compartment, the magnitudes of the stocks, or pools, of the important elements and compounds are of interest, as are the magnitudes of the flows into and out of the pools (where flows can mean physical movement or chemical transformation or a combination of the two).

For any pool that is in steady state — that is, for which inflows balance outflows so the size of the pool does not change with time — an average residence time for a compound can be calculated as the ratio of the size of the pool to the sum of the inflows or outflows.

Examination of the residence times for important pools provides quick insight into the dynamics of nutrient cycling, and, indirectly, into the probable vulnerability of specific components of nutrient cycles to disruption by human intervention. For example, a long residence time means the annual flows are small compared to the pool, which means in turn that unless human disruptions cause a very large alteration in inflow or outflow they will affect the size of the pool only very slowly. A short residence time, by contrast, means that significant changes in inflow or outflow can affect the size of the pool very quickly.

Table. Some Global Materials Flows

Pathway		Magnitude of flow (billion MT/yr)
TO OCEANS		
	River discharge (H_2O)	32,000
	Groundwater flow (H_2O)	4,000
	River-borne suspended solids (today)	18
	River-borne suspended solids (prehuman)	5
	River-borne dissolved solids	4
	Ice-sheet transport of solids	2
	Shoreline erosion	0.3
TO AND FROM ATMOSPHERE		
	Evaporation (H_2O)	526,000
	CO_2 exchange in photosynthesis	220
	CO_2 from burning fossil fuels	16
	Dust and smoke from land	1
	Volcanic gases and debris	0.1

Table. Residence Times in Some Important Nutrient Pools

Nutrient pool	Residence time (yr)
Nitrogen in atmosphere	64,000,000
Oxygen in atmosphere	7,500
Inorganic nitrogen in soil	100
Carbon in dead terrestrial organic matter	27
Carbon in living terrestrial organic matter	17
Carbon dioxide in atmosphere	5
Carbon in living oceanic organic matter	0.10
Sulfur compounds in atmosphere	0.02

CHEMISTRY OF NUTRIENT CYCLES

Two kinds of chemistry play crucial roles in the principal global nutrient cycles: the chemistry of the sedimentary cycle, associated with the very slow transformation of very large pools of material, and the chemistry of life, associated with the more rapid transformation of much smaller pools.

Both kinds of chemistry depend very heavily on the very special roles of water as a source of hydrogen (H^+) and hydroxyl (OH^-) ions for many chemical reactions, as a medium for holding other chemical compounds in solution where they can react, and as a vehicle for physical transport of compounds from place to place.

Sedimentary Cycle

Nutrients are mobilized in the sedimentary cycle when water comes in contact with rock. Rainwater is slightly acidic, having absorbed some atmospheric carbon dioxide to form weak carbonic acid:

$$H_2O + CO_2 \longrightarrow H_2CO_3$$

$$H_2CO_3 \longrightarrow H^+ + HCO_3^-$$

The pH of rainfall in areas free of industrial pollution is around 5.7. (Recall from basic chemistry that pH is the negative of the log to base 10 of the concentration of hydrogen ions in moles per liter.

The pH of distilled water is 7.0, corresponding to 10-7 moles of hydrogen ion per liter.) This slight acidity is enough to facilitate greatly the dissolution of many minerals present in exposed rock — a process called weathering. Among the most important weathering reactions is that of the common mineral feldspar (a constituent of granite):

Feldspar Carbonic acid Water

$$2KAISi_3O_8 + 2(H^+ + HCO_3^-) + H_2O \longrightarrow$$

Kaolinite Dissolved silica Dissolved potassium and bicarbonate ion

$$AI_2Si_2O_5(OH)_4 + 4ASIO_2 + 2K^+ + 2HCO_3H$$

Kaolinite is a clay; it is left behind in this process as a contribution to the formation of soil, whereas the dissolved silica, potassium ion, and bicarbonate ion are carried away in the runoff. The mobilization and removal of positive ions such as (in this case) potassium is called leaching. Note that the reaction consumes some water and that it reduces the acidity of the solution (that is, it removes hydrogen ions) as it proceeds.

These properties are characteristic of weathering reactions. Essentially the same reaction also occurs in the weathering of the aluminosilicates of sodium and calcium ($NaAlSi_3O_8$, $CaA_{12}Si_2O_8$). Other silicate minerals also weather in the same general manner to produce kaolinite or other clays.

The weathering of certain other minerals leads to their complete dissolution; no solid trace is left behind to be incorporated into soil. The most important of these minerals are the limestones, or carbonate rocks, which undergo the following reactions:

$$CaCO_3 + H^+ + HCO_3^- \longrightarrow Ca^{++} + 2HCO_3^-$$

$$CaMg(CO_3)_2 + 2(H^+ + HCO_3^-) \longrightarrow Mg^{++} + Ca^{++} + 4HCO_3^-$$

Still other weathering reactions consume atmospheric oxygen in the oxidation of iron minerals:

$$4FeSiO_3 + O_2 + 2H_2O \longrightarrow 4FeO(OH) + 4SiO_2$$

$$4Fe_3O_4 + O_2 \longrightarrow 6Fe_2O_3$$

$$4FeS_2 + 15O_2 + 8H_2O \longrightarrow 2Fe_2O_3 + 8H_2SO_4$$

Most of the products of the weathering reactions mentioned here eventually reach the sea, either as dissolved ions or as suspended solids. Between the time they are freed from rock and the time they reach the sea, of course, some of these elements have made one or more passages through the living part of the biosphere as nutrients. In the sea, too, the nutrient substances are captured from the water by living organisms, eventually to be returned to the water or to the sediments on the ocean floor.

The reverse of many of the weathering reactions also can occur in seawater in the absence of intervention by living organisms, forming solid precipitates that join the organic debris on the ocean bottom. Apparently these outputs of the major nutrient elements from the oceans roughly balance the inputs, because considerable physical evidence indicates that the salinity of seawater has been approximately constant for at least the past 200 million years.

The sedimentary cycle is finally closed by the geologic processes that is, by the processes of transformation and transportation that turn seafloor sediments into continental rocks, and by volcanic action.

Living Matter

Ecologist Edward S. Deevey, Jr., has pointed out that living matter can be regarded as a chemical compound with the empirical formula

$$H_{2960}\,O_{1480}\,C_{1480}\,N_{16}\,P_{1.8}\,S.$$

This enormous (and completely intentional) oversimplification serves both to cail attention to the chemical side of life the side concerned and, more specifically, to illustrate the relative proportions in which the six most abundant elemental constituents of living material exist.

Actually, there is substantial variation in the ratios of the chemical elements of life in different kinds of organisms, the above formula being a weighted average in which woody plants predominate. The numbers are, in any case, not known with great accuracy. The basic reactions that link the water cycle, the carbon dioxide cycle, and the oxygen cycle through living matter are photosynthesis:

$$6CO_2 + 12H_2O + \text{Solar energy} \longrightarrow + 6O_2 + C_6H_{12}O_6 + 6H_2O$$

and its opposite, respiration, or oxidative metabolism:

$$6CO_2 + C_6H_{12}O_6 \longrightarrow 6O_2 + 6H_2O + \text{heat}$$

The compound ($C_6H_{12}O_6$, or $(CH_2O)_6$, is glucose, the simplest carbohydrate produced by photosynthesis. The photosynthesis reaction is written with what appear to be six extra H_2O molecules on each side because twelve molecules of H_2O are actually broken up for each molecule of $C_6H_{12}O_6$ synthesized; radioactive-tracer experiments show that all twelve oxygen atoms from the $12H_2O$ end up in the $6O_2$, and the oxygen in the six new water molecules comes from the CO_2.

The amount of energy transformed from light into stored chemical energy by the photosynthesis reaction is 112 kilocalories (469 kilojoules) per mole of carbon fixed. The net result of the respiration reaction is to transform the same 112 kilocalories per mole of carbon from chemical energy into heat.

Photosynthesis is carried out not only by higher green plants (such as oak trees, grasses, and ferns) but also by various kinds of algae (green, brown, red, blue-green) and by certain bacteria. In the bacteria, substances other than water serve as the sources of hydrogen for the reduction of carbon. For example, the green sulfur bacteria use hydrogen sulfide instead of water, according to the equation

$$12H_2S + 6CO_2 + \text{light} \longrightarrow C_6H_{12}O_6 + 6H_2O + 12S$$

producing elemental sulfur rather than oxygen. (In fact, these bacteria are obligate anaerobes — they would be poisoned by oxygen.) The reverse reaction (respiration) consumes sulfur and carbohydrate and liberates heat, carbon dioxide, and hydrogen sulfide.

Still other photosynthesis reactions use hydrogen, obtained from H_2O or H_2S or other hydrogen donors, to reduce substances other than carbon dioxide. For example, most higher plants can reduce nitrate, instead of CO_2, in photosynthesis:

$$9H_2O + 2NO_3^- + \text{light} \longrightarrow 2NH_3 + 6H_2O + 4\tfrac{1}{2}O_2$$

and nitrogen-fixing photosynthetic organisms can reduce atmospheric nitrogen:

$$6O_2 + C_6H_{12}O_6 \longrightarrow 6CO_2 + 6H_2O + \text{heat}$$

These variations notwithstanding, the great bulk of photosynthesis on Earth consists of the reduction of carbon dioxide, with water as the hydrogen source.

The photosynthesis and respiration reactions, as we have written them here, actually represent only the end points of vastly more complicated sequences of chemical events. For example, although burning glucose in air is described essentially completely by the respiration reaction given

$$6O_2 + C_6H_{12}O_6 \longrightarrow 6CO_2 + 6H_2O + \text{heat},$$

to achieve the same result in a living cell (at a temperature far below that of combustion) requires many intermediate steps. The intermediate steps in both respiration and photosynthesis rely heavily on three kinds of organic compounds: the energy carrier ATP (adenosine triphosphate); the electron carrier NADP (nicotinamide adenine dinucleotide phosphate); and various catalysts called enzymes. (Catalysts speed up chemical reactions without themselves being consumed.)

A closer look at these compounds provides some clues to the importance of the elements nitrogen, phosphorus, and sulfur in the chemistry of life. All biochemical reactions of importance (not just photosynthesis and respiration) require enzyme catalysts if they are to proceed at significant rates.

All enzymes are *proteins*. Proteins other than enzymes also serve as hormones; as vehicles to transport oxygen; as important structural components of skin, tendons, muscle, cartilage, and bone; and in numerous other roles. All proteins are made up of building blocks called *amino acids*, and all amino acids contain the amino group $-NH_2$.

A single protein contains from fifty to some tens of thousands of amino acids, and hence at least that many nitrogen atoms. (Some amino acids contain more than one nitrogen.) The shape and stiffness of many proteins — properties essential to their functions — are governed and maintained by bonds in which sulfur plays a crucial role, so it, too, is essential to the chemistry of life.

ATP and NADP are examples of *nucleotides*, compounds consisting of a five-carbon sugar to which one or more phosphate groups ($-PO_4$) and a *nitrogen* base are attached. A nitrogen base is a ring compound containing both nitrogen and carbon in the ring structure. Nucleotides not only perform critical energy-transfer and electron-transfer functions in the chemistry of living cells, but they are also the building blocks of *nucleic acids*.

Each nucleic acid molecule contains hundreds to thousands of nucleotides, the arrangement of which is a chemical means of storing information. Nucleic acids control the synthesis of proteins and a variety of other processes in living cells.

The nucleic acid DNA (deoxyribonucleic acid) is the *genetic material* — the bearer of the coded information that is passed from parent cells and organisms to their offspring to "tell" them what to become. The nucleic acid RNA (ribonucleic acid) plays essential roles in carrying out the instructions coded in DNA molecules.

In addition to its role as a constituent of DNA and RNA, phosphorus, in the form of phosphate, is one of the principal anions (ions with a negative charge) that maintains electric neutrality in the fluids of living organisms. Along with calcium, it is also an important component in shells and bony structures. Sulfur, besides playing its role in proteins, in the form of sulfate (SO_4) is another principal anion in body fluids and cells.

The third principal anion is the chloride ion (Cl). The principal cations (positively charged ions) are sodium, potassium, calcium, and magnesium. Magnesium is an essential constituent of many enzymes and of the green pigment chlorophyll. Iron is a key component of many enzymes and of the oxygen-transporting protein hemoglobin.

Oversimplifying considerably, one can say that the quantity of living matter in a given environment is *limited by the stock of that requisite of life that is in shortest supply*. This rather obvious proposition carries the name *Liebig's law of the minimum*.

Of course, the limiting factor varies from one type of organism to another and from one environment to another. Generalizations are difficult, but it is fair to say that under many circumstances the quantity of plant material in a given environment is in fact controlled by a *limiting nutrient* (although the density of a given kind of plant is often controlled by a herbivore). In desert regions the limiting nutrient is generally water.

Under many conditions where water is not limiting, as in freshwater lakes, the limiting nutrient turns out to be phosphorus; in the open ocean it may be iron.

In some richly productive environments, nitrogen and phosphorus are jointly limiting — producing any more plant material would require supplying more of both nutrients. Obviously, the limiting nutrient in a given biological system represents a leverage point where the effects of natural or human-induced perturbations manifest themselves rapidly.

CYCLES OF THE PRINCIPAL NUTRIENTS

The cycles of many nutrients are tightly linked chemically and biologically. This is so much the case for the cycles of carbon, oxygen, and hydrogen that we consider them together. The cycles of nitrogen, phosphorus, and sulfur, on the other hand, can usefully be examined separately.

Table. Phytomass and Net Primary Production on Land and in the Oceans

	Phytomass	*Estimated net primary production*			
	(10 MT, dry weight)	*(10 MT dry organic matter/yr)*	*(10 moles carbon/yr)*	*(10 kcal/yr)*	*(10 watts)*
On Land	1840.0	107	4.0	0.48	63
In Oceans	3.9	55	2.1	0.25	33
Totals	1843.9	162	6.1	0.73	96

Carbon-Oxygen-Hydrogen

It is appropriate to begin our closer examination of the carbon-oxygen-hydrogen cycles with their intersection in the green plant, the basis of almost all life on Earth.

Carbon and Energy in Living Plants

Most living plants not only fix (store) energy by means of photosynthesis but they also use some of that stored energy to drive their own life-sustaining internal processes by means of respiration. The total rate at which a plant or a plant community stores solar energy by means of photosynthesis is called *gross photosynthesis or gross primary production* (GPP).

The figure obtained by subtracting from gross photosynthesis the amount of energy used by the plants in respiration (R) is called *net photosynthesis or net primary production* (NPP). In abbreviated form, then, the relation is GPP - R = NPP. Several interchangeable units are used in the biological literature for the measurement of these quantities: calories or kilocalories per unit of time, watts or kilowatts (different units for measuring the same thing — energy flow), grams of carbon fixed into organic compounds per unit of time, moles of carbon fixed per unit of time, or grams of dry organic matter produced per unit of time.

A wide range of estimates of global net primary production and the associated standing crop of living plants (*phytomass*) have been published. One recent review indicates that the estimates of net primary production are converging on a value around 160 billion (1.60

× 10) metric tons per year for the globe. Global phytomass was estimated in the same review as about ten times larger — 1840 billion metric tons, dry weight — and gross photosynthesis was estimated to be about twice as large as net production (that is, half of gross photosynthesis is consumed in plant respiration and half remains as net production).

The estimates for phytomass and net primary production, separated into terrestrial and oceanic components.

The roughness of these estimates deserves emphasis. The range of recent estimates of land phytomass 30 per cent larger; the range of published values of ocean phytomass is from twenty times smaller than the value to three times larger. The factor-of-60 range of informed opinion with respect to oceanic phytomass is especially remarkable.

The range in recent published estimates of oceanic net primary production is only ± 10 per cent from the value, and the range in terrestrial net production is from 15 per cent less to 60 per cent more than the value in the table. The fraction of gross primary production consumed in plant respiration is known to vary between 20 per cent and 75 per cent, depending on type of plant and region, but the global average is unlikely to differ much from 50 per cent.

Energy and Material Flow in the Carbon Cycle

For the purpose of reconciling estimates of annual production presented in grams of carbon, grams of dry organic matter, and kilocalories, one widely used convention assumes that 10 kilocalories of energy are stored per gram of fixed carbon and that dry organic matter contains 45 per cent fixed carbon by weight.

(The actual range of values for different kinds of dry organic matter extends roughly ± 15 per cent from these nominal figures.) Thus, measurements in grams or metric tons (1 MT = 106 g) of carbon can be converted to grams or metric tons of dry organic matter by multiplying by 2.2, and 1 gram of dry organic matter can be taken to represent 4.5 kilocalories of stored energy.

Useful conversions following directly from this convention are that I gram of dry organic matter per day corresponds to an average energy flow of about 0.2 watts; 1 gram of carbon per day, to about 0.5 watts; and one metric ton of dry organic matter per year, to about 600 watts. Also, 1 metric ton of dry organic matter contains 37,500 moles of carbon.

Although we shall use the convention just described, the reader may encounter others elsewhere. If the dry organic matter were pure carbohydrate — $(CH_2O)n$ — which stores 112 kilocalories per mole of carbon, then the appropriate conversions would be: 3.7 kilocalories per gram of dry organic matter, 9.3 kilocalories per gram of carbon, 0.4 gram carbon per gram of dry organic matter. [Actually, CH_2O is a good approximation of the average composition of all living matter because so much of the world's biomass is cellulose tied up in woody plants, and cellulose has nearly this formula. However, most of the plant material *produced* each year is not cellulose — remember, a big pool does not necessarily mean a big flow. The annual production of organic matter contains a higher proportion of fixed carbon and higher energy content per gram than does the cellulose-dominated, slow-toturn-over standing crop.] The rounded-off value of 4 kilocalories per gram of dry organic matter is in fairly widespread use in conjunction with the figure of 10 kilocalories per gram of fixed carbon, itself a rough average among several kinds of plant matter.

Taking the range of estimates into account, one can state several conclusions with reasonable assurance:

- The annual NPP of the oceans is only one-third to one-half that of the continents, even though the oceans cover more than 70 per cent of Earth's surface.
- The pool, or standing crop, of living plant material on land is at least several hundred times that in the oceans.
- The average turnover time (life expectancy) of oceanic plants must be in the range of days to weeks, in order for so small a phytomass to be associated with so large an annual production.
- The average turnover time of terrestrial living plants, by contrast, is in the range of a decade or two.
- Oceanic and terrestrial phytomasses are both so poorly known that man-induced or natural changes large enough to be very important in the long term could easily escape notice for some time.

In the latter regard, although there are good reasons to suppose that the activities of civilization have diminished the terrestrial phytomass significantly, some experts consider this assertion to be debatable.

Notwithstanding the uncertainties in the data, it is instructive to examine the large regional variations in plant productivity that are concealed in the global totals. Of particular interest are the facts that tropical forests account for more than 40 per cent of terrestrial NPP (although they cover only about 16 per cent of the land area) and that up- welling regions, continental shelves, reefs, and estuaries yield almost 25 per cent of oceanic NPP while accounting for less than 9 per cent of the ocean's area.

Table. Variation and Distribution of Net Primary Production

	Range of NPP (g organic matter/m/yr)	Total NPP (billion MT/yr)
Tropical Forest	1300-6000	45.3
Shrubland, Grassland, Savanna	500-3200	19.8
Temperate Forest	1200-2000	14.9
Northern Forest	600-1000	9.6
Cultivated Land	350-9400	9.1
Swamps, Lakes, Marshes	1000-1500	6.2
Desert, Tundra, Alpine Meadow	3-250	2.4

Terrestrial total		107.3
Open Ocean	50-370	42.0
Continental Shelf	370-750+	9.6
Reefs, Estuaries	1800-4900	3.6
Upwelling Regions	370-750	0.2

Oceanic total		55.4

The rate at which energy is being stored in net primary production is about 100 million megawatts, roughly two-thirds on land and one-third in the ocean. The rate of gross primary production can be assumed to be about 200 million megawatts.

The net primary production is roughly twelve times the rate at which civilization used commercial energy in 1975, where commercial energy refers to energy that is sold (for example, fossil fuels, hydropower, and nuclear power, but excluding food) as opposed to energy that is gathered by the user (for example, most wood, dung).

Very little of the solar energy that reaches Earth's surface is captured by photosynthesis. The net primary production is about 0.1 per cent of the incident solar energy at the surface-0.25 per cent of what strikes the land and 0.05 per cent of what strikes the ocean. Some of the reasons these percentages are so small are obvious.

Plants do not by any means cover all of Earth's surface; in many areas of the world the growing season lasts only part of the year; and in some circumstances, as already noted, plant growth is limited by lack of one or more critical nutrients. Under favorable conditions, on the other hand — as in a well watered and fertilized cornfield or a fertile swamp — NPP may reach 2 per cent or so of incident solar energy averaged over the growing season and as much as 6 per cent or 7 per cent of incident solar energy on the most favorable days.

These numbers still seem low. One additional reason for this is that only about one-fourth the solar energy reaching the ground is in the part of the wavelength range that stimulates photosynthesis (blue and red light). The other big reason is the large amount of water that must be evaporated through the leaves of plants to maintain the nutrient-bearing flow of water through them.

Evaporating this water typically takes 40 per cent or more of the energy that falls on a plant, and a much higher percentage of what is actually absorbed by the plant. (For production of every ton of wet weight of many crops – of which about one-fourth is dry organic matter – around 100 tons of water are evaporated.

This means 400 grams of water are evaporated, requiring about 0.6 kilocalories per gram evaporated, for every gram or 4.5 kilocalories of dry organic matter produced. The ratio in such a case is about 50 kilocalories for evaporation per kilocalorie of net primary production.)

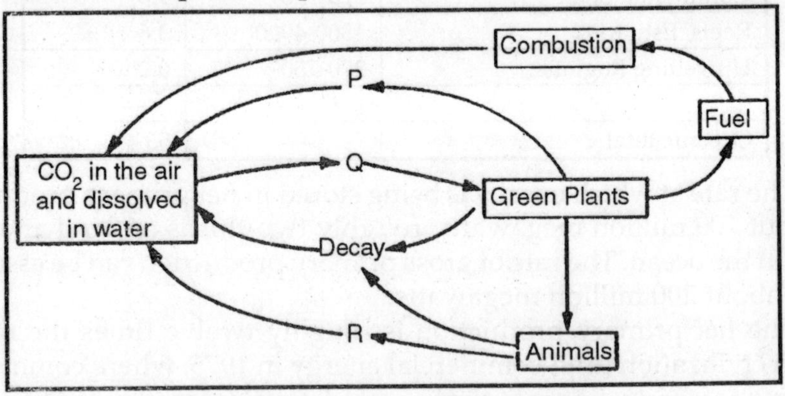

Fig. The Cycling of Carbon Through Organism.

Oxygen Balance: Metabolism by Animals and Microbes

The overall effects of the net production of plants on the cycles of carbon, oxygen, and hydrogen are as follows:

- More oxygen, O_2, is added to the environment than is removed from it;
- Less carbon dioxide, CO_2, and water, H_2O, are added to the environment than are removed from it;
- The CO_2 and H_2O subtracted from the environmental pools show up as additions to the pool of stored organic matter, CH_2O, and to the atmospheric oxygen pool.

Quantitatively, for every metric ton of dry organic matter

produced, there is a net loss from the environment of 0.6 metric tons of water and about 1.5 metric tons of CO_2, and a net gain of about 1.1 metric tons of oxygen.

Why, then, is oxygen not building up continuously in the atmosphere, while water and carbon dioxide are depleted? Because plant matter is not accumulating, but rather is approximately in steady state. It is being broken down into carbon dioxide and water again by the metabolism of animals and microbes, using up molecular oxygen just as rapidly as new plant material and oxygen are being produced.

Consider the possible fates of plant carbohydrate produced as part of net primary production. Plants either are consumed by herbivores, or they die and add their carbohydrate to the pool of dead organic matter.

Of the part consumed by herbivores, some is metabolized and thereby returned to the environmental pool of CO_2 and H_2O, some is added as excrement to the pool of dead organic matter, and some is incorporated into herbivore tissue, either to be eaten by carnivores or to be added at the herbivore's death (by causes other than being devoured) directly to the pool of dead organic matter. The same fates await the carbohydrate that reaches the carnivores.

The cycle is closed by the action of decomposers—both animals and microbes (bacteria)—which by metabolizing the fixed carbon in dead organic matter return CO and H_2O to the environmental pools.

There are some important side pathways in the carbon cycle in which the initial products of decomposition include methane and carbon monoxide (CH_4 and CO) in addition to carbon dioxide. The main reactions are:

$$2CH_2O \longrightarrow CH_4 + CO_2$$

$$2CH_2O + O_2 \longrightarrow 2CO + 2H_2O$$

The CH_4 and CO may then be oxidized in the soil or in the atmosphere according to the following reactions:

$$2CH_4 + 3O_2 \longrightarrow 2CO + 4H_2O \text{ and } 2CO + O_2 \longrightarrow 2CO_2$$

The details concealed by these simple formulas can be quite complex.

In steady state, the metabolism of carbohydrate by herbivores, carnivores, and decomposers returns to the environment exactly as much CO_2 and H_2O as the production of the carbohydrate by photosynthesis originally removed. And it consumes exactly as much oxygen as the photosynthesis produced.

There is therefore no long-run buildup (or depletion) of oxygen in the environment as a result of this cycle, unless net photosynthesis and metabolism get out of balance.

The Sequestering of Fixed Carbon

How did oxygen accumulate in the atmosphere in the first place? The biogeochemical evidence suggests that molecular oxygen in the atmosphere is of biological origin, there having been essentially none before the evolution of photosynthetic organisms, perhaps 2 billion years ago.

For oxygen produced by photosynthesis to have accumulated over the long term, part of the carbon fixed by this process must have been withdrawn from the cycle, escaping the oxidation step back to carbon dioxide. Indeed, it is precisely such a break in the carbon cycle that produced the fossil fuels.

They originated as plant material that escaped breakdown by decomposers, in many cases by being buried in swamps and bogs under conditions in which both oxygen and anaerobic decomposer organisms were absent. Hundreds of millions of years of burial under considerable pressure and heat converted these plant materials into the hydrocarbon fossil fuels—coal, petroleum, and natural gas. The oxygen released when the carbon in these fuels was first fixed by photosynthesis remained in the atmosphere.

The different ranks of coal—lignite, subbituminous, bituminous, anthracite—represent different ages (listed here from youngest to oldest) in terms of the origin of the plant material they derived from, and somewhat different chemical compositions. Peat is the partially decomposed organic matter that is the precursor of coal. Petroleum, natural gas, oil shale, and tar sands are the results of different histories of biogeochemical transformation, but all originated, like coal, with dead organic matter.

Not all of the fixed carbon that has been sequestered out of reach of natural oxidative processes is in deposits as concentrated as the fossil fuels. A much larger amount, in fact, is dispersed in sediments whose hydrocarbon content is on the order of 1 part per 1000 (1 g of hydrocarbon material per kg of sediment) or less.

Since the oxygen reservoir in the atmosphere has been built up by the sequestration of fixed carbon in various forms, could it be jeopardized by any forseeable activities of civilization that would cause the oxidation of these fixed-carbon pools?

Deforestation, for example, could lead to the oxidation of part of the standing crop of fixed carbon in living organic matter, and combustion of fossil fuels oxidizes those pools. Yet, examination of the best estimates of the sizes of the fixed-carbon pools shows little cause for concern in the short term. Removal and oxidation of the entire terrestrial biomass would deplete atmospheric oxygen by at most 0.2 per cent.

Combustion of all the conventional fossil fuels (coal, petroleum, natural gas—of which about 90 per cent is coal) thought to be recoverable ever by human effort would deplete atmospheric oxygen by 1.8 per cent. This would reduce the oxygen content of the atmosphere at sea level to what it now is at an elevation of 150 meters.

The 1975 global rate of fossil-fuel combustion would amount to a depletion of atmospheric oxygen by about 0.001 per cent per year—far below the threshold of detectability with existing measurement techniques. On the other hand, the amount of reduced carbon thought to be present in oil shales is sufficient, if it were ever totally recovered and oxidized, to consume more than half the atmospheric oxygen pool. Some observers believe that most of the oil shales are too dispersed an energy resource ever to be economically attractive.

But, in the event this view is not correct and civilization commences to use oil shale at a very high rate—say, ten times the rate at which all forms of energy were consumed worldwide in 1975—it would still take many centuries for the resulting depletion of atmospheric oxygen to become serious.

The oxygen situation is made more complicated than indicated in the foregoing paragraphs by the existence of important pools of oxygen elsewhere than in the atmosphere. It is apparent from the information that other oxygen pools must exist, because the number of moles of oxygen in the atmosphere is much smaller than the number of moles of reduced carbon in the sediments and elsewhere.

The oxygen produced when all that carbon was fixed and sequestered was tied up with the oxidation of ferrous oxide (FeO) to ferric oxide (Fe_2O_3) and sulfur to sulfate ($-SO_4$) in sedimentary rocks. It may also have been tied up in the oxidation of carbon of inorganic origin (such as carbon monoxide from volcanoes) to carbon dioxide and carbonate ($-CO_3$).

Although all these reactions are slow, they can account for the necessary enormous amounts of oxygen if they are assumed to have operated over geologic time spans. Similarly, alterations in conditions governing the reaction rates could lead to significant changes in atmospheric composition over very long periods. On shorter time scales, however, organic carbon pools and associated reactions are almost certainly more important, and they are unlikely to produce significant changes in atmospheric oxygen in less than centuries.

Carbon Dioxide

Although atmospheric oxygen depletion evidently is not a short-term threat, buildup of atmospheric carbon dioxide is. Every mole of oxygen removed from the atmosphere means a mole of carbon dioxide added. Since the number of moles of CO in the atmosphere is 700 times smaller than the number of moles of O, a change of a specified number of moles has 700 times the effect on the CO concentration as on the O, concentration.

The combustion of fossil fuel that depletes oxygen by 0.001 per cent per year augments the atmospheric CO by 0.7 per cent per year—not an insignificant increment.

Not all of the added CO remains in the atmosphere, however. Some is absorbed by the oceans, and it is conceivable that some is being taken up in an increase in biomass.

An increase in the concentration of CO_2 in the atmosphere should increase the rate of net photosynthesis (where other nutrients are not limiting), and this would produce a growing biomass if the rate of oxidative metabolism by consumers and decomposers did not increase correspondingly.

Accurate measurements of atmospheric CO_2 have been made since about 1957. They have been coupled with indirect evidence to produce a model of the rise of atmospheric CO_2 from about 280 molecular parts per million in preindustrial times to around 325 parts per million in 1975. An additional kind of evidence that can shed some light on the sources and fate of the CO_2 added to the atmosphere is the abundance of radioactive carbon-14 relative to the amount of nonradioactive carbon-12 in various pools.

Carbon-14 is produced naturally by bombardment of nitrogen by cosmic rays in the stratosphere and also by civilization when nuclear bombs are detonated in the atmosphere; it then mixes throughout the atmosphere and is incorporated, along with other carbon, into plant matter.

Since the half-life of carbon-14 is only 5770 years, the carbon-14 buried for millions of years in fossil fuels has decayed away entirely. Thus, the CO_2 added to the atmosphere by the combustion of fossil fuels contains no carbon-14, whereas CO added from pools with faster turnover does contain carbon-14.

The reduction in the $^{14}C/C$ ratio in the atmosphere because of the combustion of fossil fuels is called the Suess effect. The use of data on the Suess effect, combined with other information needed to help interpret it (such as the way the rate of natural and artificial ^{14}C production has varied over time), has allowed development of the estimate that about half of the CO ever added to the atmosphere by combustion of fossil fuels has stayed there.

The same work indicates that the land biomass should have increased between 1 per cent and 3 per cent since preindustrial times, having absorbed some of the industrial CO_2, while the oceans absorbed the rest. This change in land biomass is much too small to measure, and the exact amount is a matter of controversy.

Global Carbon Cycle

The pools and flows of the global carbon cycle. We caution the reader once more that substantial uncertainties exist about many of the numbers, as noted in detail earlier, and that further research might therefore alter the picture considerably. The pools of living and dead organic matter on land and in the oceans are assumed to be in equilibrium. Half the CO_2 added to the atmosphere by combustion of fossil fuels is assumed to dissolve in the surface layer of the oceans.

The carbon added to the atmosphere by the respiration of animals and microbes, only a few per cent is in the form of methane (CH_4) and carbon monoxide (CO), while the rest is carbon dioxide. Still, the natural sources of carbon monoxide—incomplete decomposition of organic matter and partial oxidation of methane in the atmosphere—exceed man-made carbon monoxide from fossil-fuel combustion several fold.

Methane is apparently not toxic at any concentrations that occur in natural environments, and carbon monoxide is toxic only at concentrations on the order of 100 times the global atmospheric average or more.

NITROGEN

The nitrogen cycle is chemically the most intricate of the major nutrient cycles, and it is probably the least well understood scientifically. The crucial role of nitrogen in all proteins makes this element essential to life, but most organisms can assimilate and use nitrogen only in specific chemical forms. Only a relatively few species of bacteria and algae can convert gaseous molecular nitrogen (N_2) into the more complex compounds that can be used by plants and animals. The principal usable forms are ammonia (NH_3) and nitrates ($-NO_3$), of which the latter is needed in greater quantity by most plants. The conversion of N_2 to ammonia and nitrate is called *nitrogen fixation*, and the species that can accomplish this conversion are called nitrogen-fixing organisms.

Thus, although the atmosphere represents an enormous reservoir of molecular nitrogen, the continuation of life on Earth depends absolutely on the activities of these tiny, inconspicuous organisms.

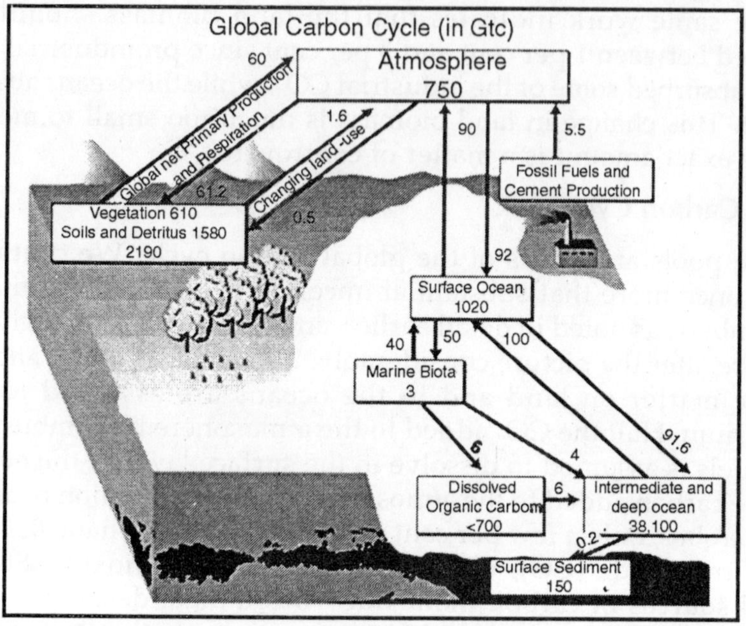

Fig. Pools and Flows in the Global Carbon Cycle.

Organisms and Nitrogen

The best known of the nitrogen-fixing organisms are the bacteria associated with the special nodules on roots of legumes, which are plants of the pea family. In a symbiotic relationship, the bacteria obtain from the plants the energy they need to carry out the nitrogen fixing reaction

$$2N_2 + 6H_2O \longrightarrow 4NH_3 + 3O_2$$

and some of the ammonia, in turn, is made available to the plants for the synthesis of amino acids, for example,

$$2NH_3 + 2H_2O + 4CO_2 \longrightarrow 2CH_2NH_2\ COOH + 3O_2$$

The capacity of such legumes as alfalfa, beans, peas, and clover to fix nitrogen in large quantities has led to their widespread use in crop-rotation schemes to replenish in the soil the nitrogen depleted by other crops.

Although nitrogen fixation by the legume-bacteria symbiosis is thought to be responsible for most natural nitrogen fixation, the fixation step is also carried out by some bacteria that have looser associations with plants, including the free-living *Azotobacter* (aerobic) and *Clostridium* (anaerobic), and by various species of blue-green algae.

Blue-green algae are apparently responsible for maintaining the fertility of rice paddies in much of Asia, which are subjected to intensive cropping without the use of nitrogen fertilizers.

The amino acids used by all organisms as the building blocks of protein are synthesized not only from ammonia but also from nitrate produced from ammonia by processes collectively called *nitrification.* The nitrification reactions yield energy, which supports the life processes of the nitrifying bacteria (*Nitrosomonas, Nitrobacter*). The actual synthesis of amino acids is carried out by bacteria (which differ widely in how many amino acids they can synthesize), by plants (which can synthesize all the amino acids needed for making proteins), and by animals (which, again, differ in the variety of amino acids they can produce).

Vertebrates—including human beings—cannot synthesize all the amino acids they need and must obtain intact in the food they eat the ones they cannot synthesize. The amino acids an animal must get from its diet are called *essential amino acids;* those it can synthesize are called *nonessential.*

Fixed nitrogen that has been incorporated into organisms returns to the soil in animal wastes and in dead organisms (microbes, plants, animals) or parts of organisms (for example, leaves). Animal wastes are rich in urea—$(NH_2)_2CO$—which is the principal product of the metabolism of proteins in many organisms. The proteins of dead organisms are broken down into amino acids and other residues by bacteria and fungi of decay.

The latter compounds—urea, amino acids, and other breakdown products of protein—are then converted into ammonia by yet another group of bacteria. This step is called *ammonification.* (The chemical reaction for the ammonification of an amino acid is just the opposite of amino acid synthesis.)

So far we have described how nitrogen is fixed by living organisms and how the fixed nitrogen is further processed by them. A logical question at this point is: How is nitrogen *unfixed*? For without a mechanism for returning nitrogen to its molecular form (N_2), even the vast reservoir of this compound that exists in the atmosphere would have been depleted long ago; much of the nitrogen would be tied up as nitrates in soil, in the oceans, and in sediments.

The mechanism that has prevented this outcome by closing the atmospheric loop of the nitrogen cycle is the action of a relatively few groups of *denitrifying bacteria* that make their living converting nitrate to nitrous oxide (N_2O) and N_2. (These transformations release energy.) The nitrous oxide is reduced to N_2 by further bacterial action, or in the atmosphere by photochemical reactions. Another set of denitrifying reactions, also carried out by bacteria, transforms nitrate to nitrite and nitrite to ammonia.

The chemistry of the main steps in the nitrogen cycle that are mediated by bacteria is summarized. For each reaction, the "energy" column indicates whether energy must be supplied to accomplish it ("in") or whether the reaction makes energy available ("out"). The part of the terrestrial nitrogen cycle directly associated with organisms is represented schematically. The oceanic cycle is substantially similar, relying both on bacteria and algae.

Chapter 8

Resources and Environment

DIMENSIONS OF THE HUMAN PREDICAMENT

Providing people with the ingredients of material well being requires physical resources - land, water, energy, and minerals — and the supporting contributions of environmental processes. Technology and social organization are the tools with which society transforms physical resources and human labour into distributed goods and services. These cultural tools are embedded in the fabric of the biological and geophysical environment; they are not independent of it.

As the number of people grows and the amounts of goods and services provided per person increase, the associated demands on resources, technology, social organization, and environmental processes become more intense and more complicated, and the interactions among these factors become increasingly consequential. It is the interactions — technology with employment, energy with environment, environment and energy with agriculture, food and energy with international relations, and so on — that generate many of the most vexing aspects of civilization's predicament in the last quarter of the twentieth century.

This book is about that predicament: about its physical underpinnings in the structure of the environment and the character of natural resources; about its human dimensions in the size, distribution, and economic condition of the world's population; about the impact of that population on the ecological systems of Earth and the impact of environmental changes on humanity; about the technology, economics, politics, and individual behaviour that have contributed to the predicament; and about the changes that might alleviate it.

THE ESSENCE OF THE PREDICAMENT

In a world inhabited in the mid- 1970s by a rapidly growing

population of more than 4 billion people, a massive and widening gap in well-being separates a rich minority from the poor majority. The one-third of the world's population that lives in the most heavily industrialized nations accounts for 85 per cent of the global personal income and a like fraction of the annual use of global resources.

The people living in the less industrialized nations must apportion the remaining 15 per cent of global income and resource use among two-thirds of the world's population. The result is an unstable prosperity for the majority of people in the DCs and frustrating, crushing poverty for the majority in the LDCs.

Millions of the poorest — especially infants and children — have starved to death every year for decades; hundreds of millions have lived constantly, often consciously, almost always helplessly on the brink of famine and epidemic disease, awaiting only some modest quirk of an environment already stretched taut — an earthquake, a flood, a drought — to push them over that edge.

The 1970s brought an apparent increase in such quirks — 1972 and 1974 were years of flood, drought, and poor harvests. World food reserves plummeted, and millions more human beings were threatened by famine. Meanwhile, the entire population continued growing at a rate that would double the number of people in the world within 40 years.

The prosperity of the DCs - awesome by comparison with the poverty of the LDCs — has been built on exploitation of the richest soils, the most accessible fossil fuels, and the most concentrated mineral deposits of the entire globe — a one-time windfall. As they now struggle to maintain and even expand their massive consumption from a resource base of declining quality, the DCs by themselves appear to be taxing technology, social organization, and the physical environment beyond what they can long sustain.

And the LDCs, as they try to follow the same path to economic development, find the bridges burned ahead of them. There will be no counterpart to the windfall of cheap resources that propelled the DCs into prosperity. A DC-style industrialization of the LDCs, based on the expensive resources that remain, is therefore probably foredoomed by enormous if not insurmountable economic and environmental obstacles.

The problems arising from this situation would be formidable even if the world were characterized by political stability, no population growth, widespread recognition of civilization's dependence on environmental processes, and a universally shared commitment to the task of closing the prosperity gap.

In the real world, characterized by deep ideological divisions and territorial disputes, rapid growth of population and faltering food production, the popular illusion that technology has freed society from dependence on the environment, and the determined adherence of DC governments to a pattern of economic growth that enlarges existing disparities rather than narrowing them, the difficulties are enormously multiplied.

Interactions

It takes water and steel to produce fuel, fuel and water to produce steel, fuel and water and steel to produce food and fibre, and so on. The higher the level of industrialization in a society, the more intimate and demanding are the interconnections among resources. Agriculture in the United States, Europe, and Japan, for example, uses far more fuel, steel, and mineral fertilizers per unit of food produced than does agriculture in India or Indonesia.

The interconnections among resources also become more intense as the quality of the resource base diminishes; the amount of fuel and metal that must be invested in securing *more* fuel and metal increases as exhaustion of rich deposits forces operations deeper, farther afield, and into leaner ores.

These tightening physical links among resources have their counterparts in economic and organizational effects. Scarcity or rising prices of one commodity generate scarcity and rising prices of others, thus contributing both directly and indirectly to inflation and often to unemployment.

Massive diversion of investment capital and technical resources to meet the crisis of the moment — attempting to compensate for lack of foresight with brute force applied too late — weakens a system elsewhere and thus promotes crises in other sectors later. Apparent solutions seized in haste and ignorance cut off options that may be sorely missed when future predicaments arise.

International aspects add to the complexity — and the dangers — of these interactions. Money pours across international boundaries, collecting in those parts of the world where rich deposits of essential resources still remain. In resource-importing DCs, the pressure to redress the balance-of-payments imbalance becomes the dominant determinant of what is exported, subverting other goals.

Resources of indeterminate ownership, such as fish stocks in oceans and seabed minerals, become the focus of international disputes and unregulated exploitation. And foreign policies bend and even reverse themselves to accommodate the perceived physical needs.

The interactions of resources, economics, and politics were displayed with compelling clarity in the worldwide petroleum squeeze of the mid-1970s. The consequences of a slowdown in the growth of petroleum production in the principal producing countries, accompanied by almost a quadrupling of the world market price, reverberated through all sectors of economic activity in DCs and LDCs alike.

The prices of gasoline, jet fuel, heating oil, and electricity soared, contributing directly to inflation. Increased demand for petroleum substitutes such as coal drove the prices of those commodities up, thereby contributing indirectly to inflation, as well. Shortages of materials and services that are particularly dependent on petroleum for their production or delivery quickly materialized, feeding inflation and unemployment still further.

The impact of the energy crisis was especially severe on the already precarious world food situation. Indeed, rising food prices, following the poor harvests of 1972, were a major factor in the decision by the Organization of Petroleum Exporting Countries (OPEC) to raise oil prices. The strong and growing dependence of agriculture on energy — especially petroleum — soon made itself felt worldwide as an important contributor to further increases in food prices.

This was especially so in developed countries a few of which are the main sources of the exportable food supplies that determine world market prices. The largest exporter, the United States, has gratefully responded to increased foreign demand for its food exports, which has helped to pay for the nation's increasingly expensive oil imports (but which has also contributed to raising domestic food prices).

This phenomenon — the need to sell food abroad to pay for oil — may have assured continued high-quality diets for nations like Japan that can still afford to buy, but it also reduced the amount of uncommitted food reserves in the DCs that would be available to alleviate famines in LDCs too poor to buy the food they needed on the world market.

The most obvious consequence of the 1973 petroleum embargo and price rise on international political affairs was one intended by the oil-producing Arab states — a sudden diminution of DC support for the Israeli position in Middle East territorial disputes. Some less obvious effects, however, may in the long run be more significant.

The United States and several other DCs export military hardware as a major source of the foreign exchange they need to pay for imported raw materials; and the intensity with which the arms exporters hustle their wares in the international market is increasing as the oil-related balance-of-payments problem worsens. The result has been to support a spiraling arms race in LDCs, which is both a pathetic diversion of funds needed there to raise the standards of living and a profoundly destabilizing force operating against world peace.

The export of nuclear reactors, likewise encouraged by the DCs' need to pay for imported raw materials, may also have disastrous effects. Although intended for production of electricity, these reactors also provide their LDC recipients with the materials needed to manufacture nuclear bombs.

India's nuclear explosion in 1974 demonstrated for any remaining doubters that the spread of reactors can mean the spread of nuclear weapons. And it must be assumed that the distribution of such weapons into more and more hands — and into some of the most politically troubled regions of the world — greatly increases the chances that they will be used.

Technology, Environment, and Well-Being

The relationship connecting technology, environment, and well-being would constitute a deep dilemma for civilization, even in the absence of the economic and political complexities just described. Simply stated, the dilemma is this: while the intelligent application of technology fosters human well-being directly, a reducible but not removable burden of environmental disruption by that technology undermines well-being.

This negative burden includes the direct effects of technology's accidents and effluents on human life and health; the direct impact of accidents, effluents, preemption of resources, and transformation of landscapes on economic goods and services; and technology's indirect adverse impact on well-being via disruption of the vital services supplied to humanity by natural ecological systems.

These free services — including, among others, the assimilation and recycling of wastes — are essential to human health and economic productivity. It is clearly possible to reach a point where the gain in well-being associated with producing more material goods does not compensate for the loss in well-being caused by the environmental damage generated by the technology that produced the goods.

Beyond that point, pursuing increased prosperity merely by intensifying technological activity is counterproductive. Many people would argue that the United States has already passed that pivotal point.

The turning point where environmental costs begin to exceed economic benefits can be pushed back somewhat by using technologies that cause the least possible environmental damage. It cannot be pushed back indefinitely, however, because *no technology can be completely free of environmental impact.*

This flat statement — implying that continued expansion of any technology will eventually lead to environmental costs exceeding its benefits — is true on fundamental physical grounds. It means that environmental constraints will ultimately limit economic production — the product of the number of people and the amount of economic goods and services that each person commands — if nothing else limits it first.

Despite increased scientific attention to environmental problems in recent years, most of the potentially serious threats are only sketchily understood. For many such problems, it cannot be stated with assurance whether serious damage is imminent, many decades away, or indeed already occurring — with the full consequences yet to manifest themselves.

Much remains to be learned, for example, about the causation of cancer and genetic damage by low concentrations of nearly ubiquitous environmental contaminants, either alone or in combination: pesticide residues, combustion products, heavy metals, plasticizers, food additives, prescription and nonprescription drugs, and innumerable others.

The most serious and imminent peril of all may well prove to be civilization's interference with the "public service" functions of environmental processes. Agricultural production in a world already on the brink of famine depends intimately on the absence of major fluctuations in climate, on the chemical balances in soil and surface water that are governed by biological and geochemical nutrient cycles, on naturally occurring organisms for pollination of crops, and on the control of potential crop pests and diseases by natural enemies and environmental conditions.

Agents of human disease and the vectors that transport those agents are also regulated in large part by climate, by environmental chemistry, and by natural enemies. Ocean fish stocks — an important source of food protein — depend critically on the biological integrity of the estuaries and onshore waters that serve as spawning grounds and nurseries.

The environmental processes that regulate climate, build and preserve soils, cycle nutrients, control pests and parasites, help to propagate crops, and maintain the quality of the ocean habitat, are therefore absolutely essential to human well-being. Unfortunately, the side effects of technology are systematically diminishing the capacity of the environment to perform these essential services at the same time that the growth of population and the desire for greater affluence per capita are creating greater demand for them.

Nor can the public-service functions of the environment be safely replaced by technology if technology destroys them. Often the foresight, scientific knowledge, and technological skill that would be required to perform this substitution just do not exist. Where they do exist, the economic cost of an operation on the needed scale is almost invariably too high; and where the economic cost at first seems acceptable, the attempt to replace environmental services with technological ones initiates a vicious circle: the side effects of the additional technology disrupt more environmental services, which must be replaced with still more technology, and so on.

THE PROSPECTS: TWO VIEWS

The foregoing brief survey of the dimensions of the human predicament suggests a discouraging outlook for the coming decades. A continuing set of interlocking shortages is likely — food, energy, raw materials — generating not only direct increases in human suffering and deprivation, but also increased political tension and (perversely) increased availability of the military wherewithal for LDCs to relieve their frustrations aggressively.

Resort to military action is possible, not only in the case of LDCs unwilling to suffer quietly, but, with equal or greater likelihood, in the case of industrial powers whose high standard of living is threatened by denial of external resources. The probability that conflicts of any origin will escalate into an exchange of nuclear weapons, moreover, can hardly fail to be greater in 1985's world of perhaps fifteen or twenty nucleararmed nations than it has been in the recent world of five.

The growth of population — very rapid in the LDCs, but not negligible in most DCs, either — will continue to compound the predicament by increasing pressure on resources, on the environment, and on human institutions. Rapid expansion of old technologies and the hasty deployment of new ones, stimulated by the pressure of more people wanting more goods and services per person, will surely lead to some major mistakes — actions whose environmental or social impacts erode well-being far more than their economic results enhance it.

This gloomy prognosis, to which a growing number of scholars and other observers reluctantly subscribes, has motivated a host of proposals for organized evasive action: population control, limitation of material consumption, redistribution of wealth, transitions to technologics that are environmentally and socially less disruptive than today's, and movement toward some kind of world government, among others.

Implementation of such action would itself have some significant economic and social costs, and it would require an unprecedented international consensus and exercise of public will to succeed.

That no such consensus is even in sight has been illustrated clearly by the diplomatic squabbling and nonperformance that have characterized major international conferences on the environment, population, and resources, such as the Stockholm conference on the environment in 1972, the Bucharest Conference on World Population in 1974, the Rome Food Conference in 1974, and the Conferences on the Law of the Sea in the early 1970s.

One reason for the lack of consensus is the existence and continuing wide appeal of a quite different view of civilization's prospects. This view holds that humanity sits on the edge of a technological golden age; that cheap energy and the vast stores of minerals available at low concentration in seawater and common rock will permit technology to produce more of everything and to do it cheaply enough that the poor can become prosperous; and that all this can be accomplished even in the face of continued population growth.

In this view — one might call it the cornucopian vision — the benefits of expanded technology almost always greatly outweigh the environmental and social costs, which are perceived as having been greatly exaggerated, anyway. The vision holds that industrial civilization is very much on the right track, and that more of the same — continued economic growth -with perhaps a little luck in avoiding a major war are all that is needed to usher in an era of permanent, worldwide prosperity.

Which view of civilization's prospects is more accurate is a question that deserves everyone's scrutiny. It cannot be decided merely by counting the "experts" who speak out on either side and then weighing their credentials. Rather, the arguments must be considered in detail — examined, dissected, subjected to the test of comparison with the evidence around us.

This is an ambitious task, for the issues encompass elements of physics, chemistry, biology, demography, economics, politics, agriculture, and a good many other fields as well. One must grapple with the arithmetic of growth, the machinery of important environmental processes, the geology of mineral resources, the potential and limitations of technology, and the sociology of change.

It is necessary to ponder the benefits and shortcomings of proposed alternatives as well as those of the status quo; and it is important to ask where burdens of proof should lie.

Does civilization risk more if the cornucopians prevail and they are wrong? Or if the pessimists prevail and they are wrong? Could an intermediate position be correct, or are perhaps even the pessimists too optimistic? What is the most prudent course in the face of uncertainty? Making such evaluations is, of course, a continuing process, subject to revision as new arguments, proposals, and evidence come to light.

Natural Processes and Human Well-Being

Prehistoric human beings earned a living from their surroundings in much the same way as many other animals did (and do): they sought out and preyed upon other organisms — edible plants and killable animals — that shared the environment, they drank water where they found it, and they took shelter in trees and caves.

The geographical distribution, size, and well-being of the human population under these circumstances were influenced very strongly by the characteristics of the natural environment — patterns of hot and cold, wet and dry, steep and flat, lush and sparse.

These patterns of climate, topography, flora, and fauna were in turn the products of millions of centuries of interaction among natural geophysical and biological processes — continental drift, mountain-building, erosion, sedimentation, the advance and retreat of glaciers, the rise and fall of the oceans, and the evolution and extinction of various kinds of organisms.

Before the advent of agriculture, most of the natural processes and systems that so strongly influenced the human condition were themselves not much influenced by what humans did; the interaction was largely a one-way street.

Fire and stone implements apparently were used even by the forerunners of *Homo sapiens* well over a million years ago, and the wielders of those early tools certainly used them to affect their local environments, both intentionally and inadvertently. But it was agriculture, which began about 10,000 years ago, that marked the sharpest transition in the capacity of *Homo sapiens* to influence the physical environment on a large scale.

From that beginning, which permitted specialization and a scale of social organization not possible among hunter-gatherers, arose a long series of technological and social developments that produced the agricultural and industrial civilizations that today virtually cover the globe. That civilizations have transformed many aspects of the physical environment is beyond dispute. Indeed, the prevalence of "man-made" environments (such as cities) and intensively managed ones (such as wheat fields) makes it all too easy to suppose that such technological environments are now the only ones that matter.

This supposition, which one could call the humanity-is-now-the-master-of-nature hypothesis, is partly responsible for the widespread underestimation of the seriousness of environmental problems. Regardless of such beliefs, civilizations and their technological environments continue to be embedded, as they always have been, in a larger nontechnological environment.

This larger environment provides the raw materials with which technology must work, and its characteristics *together* with those of technology define the limits of what is possible at any given time.

Technology and natural environmental processes together - not technology alone — have permitted the human population and its material consumption to reach their present levels. Today human beings continue to depend on the nontechnological environment for a variety of services that technology not only cannot replace, but without which many technological processes themselves would be nonfunctional or unaffordable.

The regulation of climate, the management of soil and surface water, the environmental chemistry of nutrient cycling and control of contaminants, and the regulation of pests and disease, among others. To understand the nature and extent of humanity's dependence on these services, and the degree of their vulnerability to disruption by mismanagement and overload, one must investigate in detail the character of the physical and biological processes that provide them.

"The Physical World," discussing basic geological processes, the hydrologic cycle, and climate; "Nutrient Cycles," describing the major pathways by which carbon, nitrogen, phosphorus, and other essential nutrients move from the living to the nonliving environment and back again; and "Populations and Ecosystems," an introduction to the principles of population biology and ecology essential to an understanding of the structure and functions of biological communities.

Chapter 9

Impact of Biotechnology

The Convention on Biological Diversity requires that all Member States take measures to preserve both native and agricultural biodiversity. The intrinsic value of species and ecosystems, in addition to their value as starting material for finding new products, is the basis for these measures.

The biggest threat to biodiversity is habitat destruction. The ever-increasing spread of cities and the accompanying expansion of agriculture must be held largely responsible. Humid tropical forests are particularly valuable reservoirs of biodiversity and are currently being seriously threatened.

As the human population expands, the need for food is expected to double in the next 30 years, with the ensuing threat of massive habitat destruction particularly in the less developed countries. Increasing crop productivity on the land already under cultivation would prevent or at least reduce habitat destruction.

One of several measures aimed at increasing yields is the use of better seeds, including those enhanced by modern biotechnology. Many other measures from the technical, socio-economic and political fields need to be taken at the same time in order to balance intensification and sustainability of modern agriculture.

Modern biotechnology offers new means of improving rather than threatening biodiversity. If properly tested for both risks and benefits to humans and the environment, transgenic crops are more likely to increase agricultural biodiversity and help maintain native biodiversity rather than to endanger it. Such applications need to be judged by the criteria of improved sustainability and compared to current as well as alternative farming practices.

THE ESSENCE OF BIODIVERSITY

Biodiversity is the multitude of different living beings in a particular ecosystem or on the whole earth. Biodiversity can be seen and studied at different organizational levels: genetic, organismal and ecological. It touches upon both native environments on land and sea as well as agricultural and other man-made surroundings.

Native Biodiversity

The biodiversity we observe today is the result of 3.5 billion years of evolution. Through the processes of mutation and selection all living organisms we know today, as well as those that ever lived before, developed from one single-cell micro-organism. How this first living being arose 3.5 billion years ago is still a matter for speculation.

This unitary origin explains why all organisms share the same basic chemistry: DNA is always the storage molecule of genetic information and the complex process of protein biosynthesis is virtually the same in all organisms.

Metabolic pathways are also similar in all organisms e.g., the reactions by which energy is generated or the way in which fatty acids, sugars and amino acids are made. Separate species arose when mutations between relatives no longer allowed for interbreeding, for instance after geographic or reproductive separation.

Dinosaurs are by no means the only creatures of the past that became extinct: far more organisms lived at some time on our earth than are living here today. The vast majority, probably more than 99 per cent, of species that arose on this globe, disappeared again.

This shows that evolution is an ongoing process, with species coming and going. This dynamic situation is important when discussing the conservation of species living here today. In the long-term view, there has never been any stability in life on earth - only change. However, these changes were very slow compared to the length of a human life, or even compared to the time humans have existed.

Clearly, today, with the massive amount of human interference on the globe, changes are much faster than at any other time in the last 65 million years, the point at which the trilobites and later the dinosaurs and many other creatures vanished from the surface of the globe in a relatively short time period.

As suggested by Raven in 1992, the number of species of plants, animals and eukaryotic micro-organisms is probably around 10 million today, but only 1.4 million have been characterized and given a name by scientists. There is much variation in what is known about the different groups. Virtually all of the 40 000 vertebrate animals are known and most of the 300 000 vascular plant species as well.

On the other hand, there are likely to be over a million species each of fungi and nematodes, of which only 70 000 and 13 000 have been named. There are thought to be far more than a million different insect species as well. With prokaryotes, the situation is even more extreme: about 5 000 bacteria and viruses have been named individually, yet the total number in both of these two groups may well according to Bull be in excess of one million.

Since many micro-organisms and viruses are associated with specific plants and animals, Staley considers their own biodiversity will depend on the biodiversity of their hosts, as well as the micro-organisms' own host range.

All the different species of plants and animals do not live an independent existence, but are associated in specific communities and ecosystems to form more or less stable associations. One such association is, for instance, the humid (and dry) tropical forest, which is generally thought to have the highest degree of biodiversity, with more tree species per km^2 than there are tree species in North America or in Europe, as discussed by Burslem.

Another example comes from specific types of alpine meadows or specific sorts of rivers or ponds. Biodiversity needs to be considered not only in qualitative terms of the species present, but also in quantitative terms, considering how many individuals of certain species of plants and animals are present.

With large mammals this is quite easy to determine, but it is impossible with micro-organisms. Often the number of species found in a given ecosystem is taken as a measure of the biodiversity of that system: other criteria are more difficult to apply.

We will concentrate on terrestrial biodiversity, although it is clear that streams, lakes and oceans provide habitats for a vast biodiversity of animals, plants and micro-organisms. In addition, as calculated by Naylor in 2000, they are an important source of protein as food for humans and feed for farm animals, with about 120 millions tons used per year.

Agricultural Biodiversity

In addition to biodiversity in the wild, there is the biodiversity of organisms used for farming and other human activities. In agriculture, 7 000 species of plants are used by farmers somewhere in the world, but only 30 species provide 90 per cent of our calorific intake. Intake as observed by Haywood in 2000 Within these dominant crop species, there are many hundred thousands of varieties (landraces, cultivars) adapted to local climates, farming practices, and cultural predilections like taste, colour, structure, ability to store the products etc.

Much of this large crop diversity is important for providing the initial material for breeding. However, it must be recalled that the genetic diversity found in crops is much less broad than the genetic diversity observed in plants or animals living in the wild, which points to the importance of wild species for agricultural breeding programmes. The top three crops are wheat, rice and maize (corn) with around 500 million tons annual production each.

Traditional breeding led us in to the trap of narrowing down genomes, and perhaps wisely used biotechnology could bring back at least that part of genetic diversity which enhances pest resistance and also yield.

There are many indications that mixtures of varieties of a crop or of different crops may give higher yields and be more resistant to pests and diseases than monocultures, as reported recently by Zhu for rice in China. However, even in mixed cultures, high quality, well defined varieties and pure seeds are required and the sustainability of mixed cropping related to pest management has still to be proven.

In addition, it is still not clear, whether or not, in natural, non-agricultural habitats yield is dependent on biodiversity. Based on experimental results, some researchers claim that the loss of species leads to a reduction in biomass, while others disagree, as demonstrated by Hector in 1999 and by Kaiser in 2000. There simply may not be valid correlations between biodiversity and biomass yield, in either agricultural nor non-agricultural settings.

Human Population Expansion

The one species that is still globally expanding in numbers are humans. The world population has gone up from 2.5 billion in 1950 to 6 billion today; it is expected, by the UN, to reach 8 billion in 2015 and 9 - 10 billion in 2050. Over 95 per cent of the expected population increase will be in the developing countries. In those countries, most of the population growth will occur in the cities.

The additional population will require more space to live in, more water, more energy, more food and more services. For the years 1995 - 2020 the largest relative population increase (80 per cent) is expected in sub-Saharan Africa: in absolute numbers it is expected to go from 500 to 900 million. It is not clear whether the HIV/AIDS epidemic, with an estimated 25 million infected in this region in 2000, will substantially affect population dynamics within the next 25 years, as most of the infected are of reproductive age, as pointed out by Cohen in 2000.

It is important to realise that the FAO estimates that there are over 800 million people globally who don't have enough to eat today, and it is imperative that more food be produced for them where it is needed, namely in the developing countries. As pointed out by Lipton in 1999, Poverty reduction in rural, agricultural areas of developing countries is also likely to have an indirect effect in favour of the urban poor. Improved seeds will have a positive effect, but many other factors such as education, access to microcredit, reduction in farm subsidies in the North etc will be at least as important.

INTERNATIONAL AGREEMENTS

In view of the importance of biodiversity for the future of mankind, several international agreements have been reached. Since this has occurred only in the last few years, the long term impact of these agreements cannot yet be estimated.

The Convention on Biological Diversity (CBD)

Recognizing that the biodiversity of organisms in the wild should be maintained both for their own intrinsic value, but also on practical grounds, the United Nations prepared the Convention on Biological Diversity (CBD) and succeeded in having it adopted in 1992. It entered into force in 1993. This is the first time that a large majority of States, have agreed to a legally binding instrument for biodiversity conservation and the sustainable use of biological resources.

A radical change brought about by the CBD is the recognition that States have a sovereign right over biodiversity within their own territory: previously organisms were considered to be the common heritage of mankind. Living organisms or their products may, under the terms of the CBD, only be removed from a country under mutually agreed conditions.

The CBD is a comprehensive approach to biodiversity conservation of both wild and domesticated species. It aims at conservation at the genetic, species and ecosystem levels. As reviewed by Buhenne-Guilmin, action is delegated to the national level obliging States to assess biodiversity, enact legislation for its conservation in situ and ex situ, and to enforce legislation within national boundaries.

The field of biotechnology is particularly affected by of the CBD, since they require a fair and equitable sharing of benefits derived from the use of genetic resources. This includes providing facilities and financial means for technology transfer and open access to scientific and technical information.

The sovereignty over biological resources means that noone can remove specimens of plants, animals or micro-organisms from a country without the prior consent of that country. One example of a joint effort in "bioprospecting" is a search for specific active ingredients of plants by the Merck Company in the tropical forests of Costa Rica. This brought the country 2 million dollars over a five year period, in addition to the potential for royalties, should profitable products emerge. A small number of similar agreements have been undertaken elsewhere in the world.

The regulations of the CBD have only been in operation for a few years. It is too early to assess their long-term effects. As far as biotechnology and "bioprospecting" is concerned, it will take more time to establish smooth administrative procedures to allow simple routine implementation of close collaborations. This system will only be perpetuated if both the national authorities from countries rich in biodiversity, and pharmaceutical or other companies see the mutual advantages to be gained from such collaboration.

One of the real obstacles to this are the exorbitantly high costs of drug testing that needs to be done to meet strict and justified regulation before marketing. The expectations of some LDCs to make rapid earnings may have been too optimistic. The search for natural, highly active pharmaceuticals in wild plants may often be more cumbersome than laboratory searches using genomics, rational drug design and combinatorial chemistry..

The Cartagena Protocol on Biosafety

The CBD provided a basis for developing and formulating a further international agreement, namely one primarily regulating the transboundary movement of living GMOs. After much debate in 1999, the new protocol was agreed upon in early 2000 under the name of the "Cartagena Protocol on Biosafety".

The "Intergovernmental Committee for the Cartagena Protocol on Biosafety" had its first meeting in Montpellier in December 2000. It paved the way for launching a pilot phase of the "Biosafety Clearing House", a centre for information exchange, as explained on the internet.

The Protocol is a worldwide regulation for the transfer, handling and use of GMOs that may have an adverse effect on biodiversity, also taking into account risks to human health and focusing on transboundary movements. It makes explicit reference to the precautionary approach. It establishes an Advance Informed Agreement (AIA) procedure for imports of GMOs intended for introduction into the environment and an alternative procedure for mass movements of GMOs intended for food, feed and for processing (commodities).

The permit for transboundary movement will or will not be issued on the basis of a risk assessment procedure performed by the national competent authority. The Protocol does not pertain to pharmaceuticals or other non-living products made by genetic modification.

In practice, the Protocol will be most important for the import of transgenic seeds. It requires a risk assessment by the national authorities and allows countries to reject GMOs. The protocol specifies that "lack of scientific certainty due to insufficient scientific information and knowledge regarding the extent of the potential adverse effects... shall not prevent the party from taking a decision".

This may or may not be in agreement with the WTO rules, but will need to be tested in the courts. If there are disputes between the interpretation of the Protocol and the WTO regulations, which allow for trade barriers virtually only if there are scientific reasons to do so, the outcome of such disputes will depend to a considerable degree on the quality of scientific data used for assessing the benefits and risks of GMOs.

The Protocol is likely to become operational in the year 2002, after ratification by 50 nations. Early in 2001 it had been signed by 86 nations and ratified by two. The Protocol will, as pointed out by Mahoney, be a challenge to the scientific community to provide solid scientific data to convince national authorities of the benefits of transgenic crops and their relatively low and manageable risks. It is far too early to make any assessment of the Protocol's effect.

Hopefully it will allow (and not prevent) the planting of GM crops that may have real advantages for developing countries, while at the same time minimizing their risks to humans and their environment. The Protocol may contribute to a better public understanding of what transgenic crops are, but it may on the other hand, by treating GMO crops as a special group, increase fears of a negative impact, even in the absence of any science-based data showing such an impact. It will be difficult to move forward with the Precautionary Approach, if it is taken as a stringent principle.

LOSS OF BIODIVERSITY AND CONSERVATION

Losses of biodiversity are undoubtedly occurring in many parts of the globe, often at a rapid pace. These losses require countermeasures such as an increased effort towards conservation by many different means.

Reduction of Biodiversity

The loss of biodiversity can be measured by a loss of individual species, groups of species or decreases in numbers of individual organisms. In a given location the loss will often reflect a degradation or a destruction of a whole ecosystem. Recently the Subsidiary Body on Scientific, Technical and Technological Advice (SBSTTA) of the CBD ranked the priority of threats to global biodiversity in the following manner: first comes habitat loss (most of it through the expansion of cultivated land), second comes the introduction of exotic species.

Habitat loss comes not only from taking more land under the plough, but also from expanding cities and road building. In addition, habitats can be damaged by flooding, lack of water, climate changes, salination etc., all of which phenomena that may be both natural or man-made.

Since tropical humid forests are particularly rich in biodiversity, their destruction is disproportionately damaging to biodiversity. It is estimated by Pimm and Raven that of the original 16 million km^2 of these forests known a century ago, only half are left, with about one million km^2 being destroyed every 5 to 10 years. Burning and selective logging may damage an even greater area.

Biodiversity is not homogeneously distributed over the humid tropical forests, rather there are hotspots with a particularly high level of biodiversity. These hotspots are of particular interest for the implementation of conservation measures.

The second most important reason for loss of biodiversity is invasion by exotic plants and animals. Knowingly, or unknowingly, imported plant species threaten the native ones by being highly competitive and often by lacking local predators, such as insects or birds.

One of the most extreme examples is seen in the pampas of Argentina, a flat grassland with a moderate climate, from which nearly all the native grasses have disappeared and have been replaced by European plants. This invasion was brought about by European farmers who introduced animals and crops, in addition to accidentally spreading many different weeds.

This phenomenon was already noted in 1833 by Charles Darwin. Today, still, droves of gardeners transport seeds all over the globe and never think of the possible threat to biodiversity, as suggested by Ammann in 1997. It has been estimated that one in ten imported plants may spread in a modest way and that one in a hundred may turn into a nuisance weed.

Even in today's Europe, invasion by exotics may threaten ecosystems. In the Ticino region of Southern Switzerland Robinia pseudoacacia, a native of North America, is displacing chestnut and oak trees, whilst in the Northern regions of the country Solidago canadensis is replacing native irises in swampy areas. Islands are particularly threatened by invaders, as is well documented for Hawaii, New Zealand and the Galapagos Islands.

For North America, it has been estimated that the damage caused by exotics amounts to137 billion dollars a year. Although such calculations are fraught with uncertainties, there is no doubt that the costs of exotics are tremendous.

Exotic biological control agents are often introduced to agricultural ecosystems on purpose, in order to control pests or weeds without resorting to chemical control agents. Whilst there are some success stories, Strong pointed out in 2000 that such systems may also go wrong. One example is the introduction of the seven-spot ladybird, which was intended to fight the Russian wheat aphid.

The consequence, however, was the disappearance of the native ladybirds, for which the seven-spot import was a competitor and an actual predator. Another example is the decimation of the large American moths, which are killed by European Compsilura flies, introduced nearly a century ago to control the gypsy moth. Field experiments recently done by Jensen showed that caterpillars of the American moth Cecropia were killed by massive infestations of Compsilura maggots.

It cannot be said today, on the basis of experimental evidence, whether transgenic plants are specifically prone to spreading in the long term. However, one would not expect this to be the case unless the transgenic plant had an increased fitness. There is no good argument why crops that have for centuries depended for survival on human care should become weeds, just because of the addition of one or a few well characterized genes, in addition to the many thousands of genes they already carry.

However, this issue needs to be carefully studied on a case by case basis, keeping in mind that the absence of a negative effect can never be proven with absolute certainty. The results of a fairly long-term study of the performance of transgenic crops in natural habitats were recently presented by Crawley in 2001.

Four different crops (oilseed rape, potato, maize and sugar beet) were grown in 12 different habitats and monitored over a period of 10 years. In no case were transgenic plants found to be more invasive or more persistent than their conventional counterparts, in agreement with the general hypothesis put forward above.

Conservation Strategies

Conservation may be in situ or ex situ, either in the natural or semi-natural habitat, or in some purpose-built environment. The choice of one or the other technique, or a combination of both, will depend on the particular case. In situ conservation will involve the maintenance and protection of natural habitats, while botanical gardens and seed banks are used for the ex situ conservation. Both of the latter require precise knowledge of taxonomy.

Conserving a substantial, but selected fraction, of the humid tropical forests would still allow half or so of their indigenous species to be preserved. This would require a selection of the most appropriate areas, called the hot spots.

Protecting huge tracts of land will pose major socio-economic and political problems. It has been asked how forests destined to be protected from human encroachment can be kept free of hungry people in search of potential farmland.

Jennings thinks that a viable strategy may be to find a sustainable livelihood for rural populations in connection with conserving tropical humid forests. Policing alone will not be successful over vast territories, as seen today in the war on drugs in South America and Asia.

Today, conservation also embraces various components of agro-biodiversity like crop varieties, land races, semi-domesticates and crop relatives. The role of indigenous communities in maintaining agro-biodiversity is stressed by the Global Biodiversity Assessment and the Leipzig Plan of Action, two recently concluded international agreements.

APPLICATIONS OF BIOTECHNOLOGY AND ITS EFFECT ON BIODIVERSITY

The methods of biotechnology can be applied to the study of virtually any biological phenomenon and will in some cases have practical applications for maintaining biodiversity. Conversely, threats to biodiversity by biotechnology also need to be considered.

Biotechnology for the Acquisition of Knowledge

There are two quite different applications of biotechnology, or of molecular biology, that are relevant. The first is the use of biotechnology as a tool for acquiring knowledge, whilst the second is the use of biotechnology to directly intervene in plant and animal breeding, in particular to transfer genetic information from one sort of organism to a particular crop, or to a farm animal to make it transgenic.

Today, biological research can hardly be conducted without using biotechnology in one way or another. Taxonomy uses molecular markers to identify individual strains of organisms or to identify species, much in the same way as in forensic medicine to identify criminals. This is useful for ex situ conservation of plants and micro-organisms.

In seedbanks, genetic fingerprints are used to establish the origin of a seed or the relatedness of one plant variety to another. A rational classification of most micro-organisms has only become possible with these biotechnological methods. This is important to the many collections of micro-organisms that exist around the world.

Biotechnology has also proven useful for following genetic markers in plant and animal breeding. Here, animal or plant varieties are crossed by conventional, sexual means. By analyzing a few cells of the newly born calf or of the newly sprouted crop, one can predict some of the expected properties of the progeny, by looking at the presence or absence of certain forms of genes.

This enables one to predict a phenotypic property, which will only show up later in life, for instance certain characteristics of a cow's milk or the crop's expected resistance to an infectious plant disease. Using in vitro fertilization of animals, the laboratory test can be done even before the embryo is implanted. This is called pre-implantation diagnostics.

These applications of biotechnology are in general not controversial. Here, the public perception is similar to the application of modern biotechnology in medicine, for instance to industrially produced pharmaceuticals, vaccines and diagnostics. It may be noted that the worldwide sales of biotechnology products in 1999 was worth around 13 billion dollars in the medical field and only 1.6 billion in agriculture.

The availability of genome sequences will be a boost to research. The first two complete plant genome sequences determined were those of Arabidopsis and rice. The 120 million base pairs (MBP) of the small brassica Arabidopsis were sequenced by an international academic consortium and the data made public.

The 430 MBP sequence of rice was completed only a few weeks later by an industrial group lead by Syngenta, and will be available by contract to other researchers. Syngenta intends to make the data available free of charge for research directly benefiting subsistence farmers. The public sector sequencing of rice through an international consortium is expected to be completed in 2004.

It will hopefully become common practice for companies to make their basic discoveries publicly available, to everyone's benefit. The Monsanto company has also opened up some of its rice sequencing data.

DIRECT GENE TRANSFER TO CROPS AND FARM ANIMALS

Since all genes consist of DNA, and the information in this DNA molecule is read in the same way in all organisms in order to make proteins, it is in principle possible to take any (single) gene from any organism and transfer it into any other organism so that the recipient produces a protein normally only made in the donor. The resulting organism is called transgenic.

From the time this simple strategy was devised, it took molecular biologists about twenty years until the first transgenic plants were made in 1985. Ten years later, the first transgenic crop appeared in supermarkets in the USA, the "FlavrSavr" tomato. In 2000 there were, worldwide, about 45 million hectares planted with commercial transgenic crops.

Most transgenic crops planted commercially in 2000 were in the US, Canada, Argentina, with smaller amounts in China, Australia, South Africa, México and Spain. Soyabean and corn ranked first and second, making up 57 per cent and 22 per cent of the total area planted with GMOs. Cotton and canola accounted for about 5.3 and 2.8 million ha each, whilst only small areas of transgenic potato, squash and papaya were grown commercially.

With regards to the genetically-modified traits, herbicide tolerance was dominant with 74 per cent, while insect resistance was 19 per cent. According to the ISAAA, the amount of virus resistant crops was quite small. Compared to 1999 there was an increase of 10 per cent in the area planted with GMOs. For corn there was a decrease, presumably because of an anti-GMO-wave that started in Europe in early 1999; soya and cotton showed an increase.

The reason that US farmers have adopted the transgenic crops surprisingly quickly is because of the economic benefits they offer. In most surveys done by different researchers in different parts of the US, the yields were the same or somewhat higher with the new seeds The most noticeable difference to the farmers was the saving on herbicides.

The US National Centre for Food and Agricultural Policy cited an annual saving of US$ 220 million to soyabean farmers. It was also found that the new crops needed less frequent sprayings and allowed "no till" management. These benefits mostly offset the initial higher cost of the transgenic seeds, although farm profits, with or without modern biotechnology, vary a great deal from year to year and region to region. Clearly these economic considerations only hold for countries with economic structures similar to those of the US, not to developing countries. It is important to remember that a large number of transgenic crops are still in the development stage and will only come onto the market in a few years from now. They are likely to show benefits for the consumers and some may be of particular interest to farmers in tropical countries.

Two rice varieties, with anticipated consumer benefits, are those containing Vitamin A or an increased level of iron in the product, which were developed by Potrykus and Beyer last year. Despite traditional preventative measures (distribution of free vitamin A, encouragement to eat more fruit and vegetables), worldwide there are 130 million young people who are vitamin A-deficient, one to two million die annually as a consequence of vitamin A-deficiency and 500 000 turn irreversibly blind every year.

A bowl of 300 g of this cooked rice is thought to be enough to overcome the vitamin A-deficiency to a significant degree. Similarly iron-deficiency, particularly prevalent in pregnant women, can potentially be alleviated by rice containing an increased amount of iron in its endosperm. Such rice varieties have been successfully developed in the laboratory, but are far from commercialization, for both scientific and political reasons.

For farmers in developing countries, the following GMOs may be of interest:

- Virus-resistant cassava.
- Virus-resistant sweet potatoes.
- Virus-resistant papaya (already on the market in Hawaii).
- Rice with an increased rate of photosynthesis, and therefore with a potential yield increase of up to 25 per cent.
- Rice with increased salt tolerance.
- Diverse varieties that are partially aluminium-resistant and have the potential to grow in degraded tropical soils.
- Diverse crops that are more drought-resistant than the usual varieties.

All of these and many more crops have been proven to work, in principle, in laboratory and glasshouse trials. The practical benefits and risks of the crops need to be assayed in the field and their products scrutinized, like any other novel food.

Several lines of transgenic farm animals have been produced, but none have been made commercial. Some lines are made for the pharmaceutical industry to produce drugs in their milk. Others may show improved resistance towards certain infections. Transgenic salmon that grow faster than normal have been developed and have roused considerable concern amongst ecologists. As pointed out by Reichhardt, many environmental issues still need to be clarified in this context. However, it is clear that the transgenic crops that have been commercialized so far have not been seen to have done any harm to either the environment or consumers.

NATIVE BIODIVERSITY AND BIOTECHNOLOGY

Biodiversity in the wild has been massively reduced in the industrialized countries over a long period of time, in Europe, for example, over several millennia. Hardly any ecosystem is the same here as it was before humans started to clear forests and develop farming. When we look out of the window we see houses, roads and meadows, whilst three thousand years ago the area was covered in beech and oak forests.

Even Europe's forests are more like manicured gardens than virgin forests, despite the mystical "naturalness" attributed to our forests, at least in Germanic countries. North America still has far more native, untouched ecosystems, either in the form of protected areas or in less hospitable regions like the North of Canada. Biodiversity has already diminished on a massive scale in the industrialized countries.

Despite many conservation efforts, about half of the tropical humid forests have already been destroyed. How then can the rest be preserved, given that the regional population is increasing rapidly, as for instance in sub-Saharan Africa, where a near doubling of the population is expected in the next 25 years?

Already today, the per capita food supply in sub-Saharan countries is only 2100 kilocalories compared to 3500 kilocalories in the industrialized countries, so there will be a huge pressure to provide more food. Even if some food is imported, the vast majority has to be produced regionally or nationally, both for reasons of economy and ecology (e.g., poverty reduction, transport and distribution costs).

Yields of cereals have gone up very considerably in the last forty years. In the developing countries, this is primarily a result of the Green Revolution. However, the annual growth increases in cereal yields have slowed down from about 3 per cent to 1 per cent per year as shown by Pinstrup-Andersen and his collaborators in 2000.

For the developing countries they were 2.8 per cent in 1967-1982, 1.9 per cent in 1982 - 1994 and only 1.2 per cent in 1993 - 2020. These lower yield increases of recent years mean that productivity will probably not keep up with demand in the developing countries. The consequence on biodiversity is devastating and means that more land will be required for farming.

This land will primarily come from areas with high native biodiversity, in particular the aforementioned tropical humid and dry forests or from marginal land. Whilst the green revolution has also had negative consequences, such as salination, excessive water use and soil degradation, the increased productivity it achieved allowed the maintenance of large tracts of native untouched land not used for farming.

Conway concludes that the single most promising way to avoid habitat destruction is to increase farm yields in a process that has been called the second green revolution. Several components will be required to increase productivity: better training and education of farmers (in particular women), more favorable economic and political climate, availability of microcredits etc. In addition, technical contributions will also be necessary.

One such contribution is improved seed, produced either by traditional crop breeding or by modern biotechnology. There will have to be more reliance on the latter, since traditional breeding seems to have reached a yield plateau. So agricultural biotechnology, which is viewed controversially in the public debate, may contribute markedly to conserve biodiversity by preventing the appropriation of native biodiversity-rich land for farming purposes.

It should be noted, however, that this technology - like all others - is no panacea. Pinstup-Andersen and Cohen believe that each application needs to be studied carefully on a case-by-case basis, like any other new technology.

A valid concern is the possible effect of Bt-crops and similar plants on non-target insects. The Bt-crops contain a gene coding for an insecticidal protein originally produced by the soil bacterium Bacillus thuringensis. They were developed to make the plants resistant to a particular, highly damaging pest and have been quite successful in reducing pesticide input when infestation rates are high.

In laboratory studies Losey showed in 1999 that the pollen from Bt-corn could kill larvae of the Monarch butterfly when a large amount of pollen was sprayed on the larvae's favourite food plant, milkweed. Subsequent field studies by Sears showed that the Bt-corn caused little or no damage to the Monarch in real agricultural settings. This shows that the impact of transgenic crops on non-target organisms cannot be studied solely in the laboratory, but also requires farmland experimentation.

A more limited concern, that largely touches Northern Europe, is the conservation of native plants and animals, in particular birds, in farmed areas. The birds' habitats are fields, hedges, roadsides and fallow land where they depend for food on insects and seeds produced by weeds in or near the crops. These seeds are particularly important in the winter months. Computer models by Watkinson suggest that more intense weed control measures may lead to smaller amounts of seeds being available to birds.

This effect seems plausible under certain conditions, but depends on weed management regimes rather than on the presence or absence of transgenic plants, and therefore is not an issue of biotechnology. Herbicide tolerant beets may allow farmers to tolerate weeds for a longer time and fight them only after sowing.

This is made possible by a post-emergence herbicide treatment. More efficient weed management may also make it possible to set aside more land. If the lack of food results in a reduction in the bird population, this may lead to an increased number of harmful insects in the fields.

It must be remembered that farming the land serves quite different purposes, particularly in Northern Europe. The primary goal is obviously the production of food, but secondary goals, such as conservation of biodiversity and giving city dwellers opportunities for outdoor activities, are also important. For the latter purpose, setting aside more farmland would be helpful. In order to do this, political will and financial incentives are a prerequisite.

AGRICULTURAL BIODIVERSITY AND BIOTECHNOLOGY

In addition to wild plants, old landraces might be threatened by transgenic crops. It should be remembered that vertical gene transfer by pollen has always occurred between different old landraces and between different new varieties of crops. Despite this, varieties of apples or cereals have been stable over many years and specific traits have not disappeared. Pollen has always flown.

What has become far more precise is the method of analysis. Thanks to gene probes and to GMOs, it has become much easier to follow gene flow, since one is no longer dependent on visible traits, but can follow a specific gene. Nevertheless, it is important to preserve landraces and native relatives of crops for their intrinsic value as well as for having starting material for future crop breeding.

Can newly introduced transgenic crops transfer genes vertically to native wild plants and thereby change important characteristics of the wild plants? Vertical gene transfer between cultivars and wild plants has always occurred within the limits of species, if the two types of plants were in close proximity and flowered at the same time. No new problems can be expected from transgenic plants, except if the gene transferred from the GMO to the wild plant significantly increased the fitness of the recipient.

This seems rather unlikely a priori, but needs to be studied experimentally, as suggested by Ammann and collaborators in 1999, both in the laboratory and the field. Herbicide resistance transfer from a herbicide tolerant, transgenic crop to a close relative can occur in the field, as has been shown for canola by Mikkelsen in Denmark. However, this would be of major significance only if the recipient weed was controlled by this herbicide in this farm setting.

Small farmers in many developing countries use a remarkable number of races of many different crops. These are often well adapted to the local climate and topography, and are used to produce foods for different cultural purposes. In the Andes region of South America, dozens of different varieties of potatoes are grown, often side-by-side on small plots.

In Europe, different varieties of potatoes are used for the industrial production of chips, and for preparing rösti in the home. Will the traditional races disappear when, and if, transgenic crops are introduced by farmers in developing countries? To judge from the past in Europe, the answer will be largely yes, not because of any biological hazard emanating from the GMOs, but because farmers need to produce their products and sell them economically.

Many old varieties of apples have disappeared in Europe because of the preferences of the food retailers and consumers, who buy just a few varieties.

The rapid consolidation in the global seed market is also a concern for the maintenance of agricultural biodiversity: a decreasing number of companies control the global seed and agrochemical markets. Anti-trust legislation needs to prevent the formation of overly powerful monopolies.

This concern is not a consequence of the introduction of agricultural biotechnology, but a consequence of globalization. New technologies have, not only in biotechnology, but also in many other areas, made large companies more powerful and more visible. It may be possible that modern breeding methods based on molecular genetics will per se reverse the trend of variety losses, since the insight in genomics may bring back respect for races and the need to conserve them.

SOCIAL CONSEQUENCES

The introduction and spread of new technologies generally have social consequences with winners and losers. For biotechnology, this has led to intense public debate across many different facets, among them ethical, economic, legal and emotional.

Economic Considerations

One may wonder under what economic conditions biotechnology will benefit agricultural productivity, prevent the expansion of farmland and thereby help the preservation of biodiversity. It is worth remembering that the green revolution, which improved wheat and rice production in Asia, was initiated through work done in the public sector, namely in the Philippines at the International Rice Research Institute (IRRI) and in México at the Centro Internacional de Mejoramiento de Maiz y Trigo (CIMMYT), both institutions which belong to the Consultative Group of International Agricultural Research (CGIAR). Today, innovative research in agricultural biotechnology is largely done by a few large companies, with smaller contributions from the public sector.

That companies strive for intellectual property rights (IPR) through patenting or other protective mechanisms is understandable and largely unavoidable, if innovation is to be maintained. However, provisions need to be made so that the agricultural research institutions of developing countries can obtain easy access to the patented materials and procedures as well as to the knowledge they need in order to improve the position of poorer farmers, as proposed for instance by IRRI.

A very limited number of transfers of specific patents have already occurred, such as from Monsanto to the Kenyan Agricultural Research Institute for producing virus-resistant sweet potatoes or another to IRRI for the promoter genes used in golden rice. This approach, including the possibility of forced licensing, is more practicable than wanting to overthrow the world's patenting system with the slogan "No patents on life".

The patenting system does, however, need to be adapted in order to do justice to the complexity of living organisms. It may be noted in passing that the first patent granted for a living organism was given to Louis Pasteur in 1873 for a specific type of yeast.

A new issue arising in the field of IPR is the attempt to protect traditional, indigenous knowledge and preserve biodiversity through new international legislation. This legal instrument is still in an embryonic state of development.

At the level of the individual human being, the lack of food appears to be, and is, largely a problem of distribution: there is not sufficient food available where it is needed. In addition, the poor simply do not have the money to buy sufficient food. Since in developing countries most of the poor still live on the land, it is there that more income has to be generated, so that there is not only enough for survival, but a surplus which can be sold on the market. Increasing agricultural productivity is the most effective way to do this and should, according to economists, embrace modern biotechnology as an important component.

ETHICAL CONSIDERATIONS

The issues involved in the interaction between biodiversity and biotechnology have far-reaching consequences and need to be subject to an open and knowledge-based dialogue in society. The heated public debate, seen primarily in Europe in the last couple of years, and which in some circles led to near hysteria, is according to Leisinger not sufficient to solve the underlying problems.

The dialogue needs to include many different stakeholders, including farmers of developing countries, diverse scientists, policy makers and communicators. Cultural values involved in farming and food production need to be taken into consideration, just as much as the emotional side of eating and drinking.

A significant aspect of ethical behaviour is openness. Transparent information is required both from scientists in non-commercial settings as well as from industry. Although competition from fellow researchers and from industries does not allow the immediate divulgence of all research results, both groups should make concessions to the interested public and to policy makers. This should be done by scientists personally and would help improve mutual confidence.

Several large projects in biotechnology like HUGO, the Human Genome Organization, have an "ELSI" component, dealing with ethical, legal and societal implications of biotechnology. This shows that some scientists realise that science cannot be seen as a human activity taking place in a void, without any connection to social and political realities.

Ethicists stress that it is the responsibility of scientists to be actively concerned with these issues and that they should take special responsibility in communicating with non-specialists to explain what they know as well as what they don't know. Matrices of criteria have been set up for instance by Mepham to make it possible to ask relevant questions, such as who and what may be affected by agricultural biotechnology, and what properties like well-being, autonomy or justice may or may not be infringed upon.

The questions put forward by ethicists need to answered on the basis of concrete scientific knowledge, using comparisons with other, established agricultural practices. Ethicists will have critical and thought provoking questions. However, their answers and attitudes may vary considerably. If they are positive as Comstock, ethicists will suggest pursuing research and application with appropriate caution. There is no easy answer to what the word "appropriate" means, beyond answering questions on a case-by-case basis to balance the benefits and risks. Actively shaping the future and attempting to solve problems always involves risk.

It will be important to see ethics not as a stable set of principles, but rather as a discursive process, which will be able to cope with the growing speed of new developments. The Nuffield Council of Bioethics recently came to the conclusion that "The moral imperative for making GM crops readily and economically available ... is compelling", if well-informed governments of developing countries want to introduce them. This imperative should be taken seriously and treated with urgency: every day 20,000 children die of malnutrition and 30,000 ha of humid tropical forest are destroyed.

ENVIRONMENTAL BIOTECHNOLOGY

Environmental biotechnology is when biotechnology is applied to and used to study the natural environment. Environmental biotechnology could also imply that one try to harness biological process for commercial uses and exploitation.

The International Society for Environmental Biotechnology defines environmental biotechnology as "the development, use and regulation of biological systems for remediation of contaminated environments (land, air, water), and for environment-friendly processes (green manufacturing technologies and sustainable development)".

Significance Towards Industrial Biotechnology

Consider an environment in which pollution of a particular type is maximum. Let us consider the effluents of a starch industry which has mixed up with a local water body like a lake or pond. We find huge deposits of starch which are not so easily taken up for degradation by micro-organisms except for a few exemptions. We isolate a few micro-organisms from the polluted site and scan for any significant changes in their genome like mutations or evolutions. The modified genes are then identified.

This is done because, the isolate would have adapted itself to degrade/utilize the starch better than other microbes of the same genus. Thus, the resultant genes are cloned onto industrially significant micro-organisms and are used for more economically significant processess like in pharmaceutical industry, fermentations...etc.

Similar situations can be elucitated like in the case of oil spills in the oceans which require cleanup, microbes isolated from oil rich environments like oil wells, oil transfer pipelines...etc have been found having the potential to degrade oil or use it as an energy source. Thus they serve as a remedy to oil spills.

Still another elucidation would be in the case of microbes isolated from pesticide rich soils These would be capable of utilizing the pesticides as energy source and hence when mixed along with bio-fertilizers, would serve as excellent insurance against increased pesticide-toxicity levels in agricultural platform.

But the counter argument would be that whether these newly introduced microorganisms would create an imbalance in the environment concerned. The mutual harmony in which the organisms in that particular environment existed may have to face alteration and we should be extremely careful so as to not disturb the mutual relationships already existing in the environment to which we are introducing the newly discovered and cloned microorganisms.

Analysis of both the benefits and the disadvantages would pave way for an improvised version of environmental biotechnology. After all it is the environment that we strive to protect.

Chapter 10

Ecology of Microbial Beds

The artificial environment of a sewage bacteria bed has been successfully invaded by truly aquatic micro-organisms and some moisture loving higher organisms, and it is convenient to distinguish between the micro-organisms or 'film' and the higher forms of life, the grazers, although as will be appreciated, some organisms of the film are also grazers in the strictest sense.

ORGANISMS OF THE FILM

BACTERIA

Bacteria active in bacteria beds have attracted little attention; according to Wattie bacteria found in filter slimes are zöogleal bacteria of one group closely related to those in activated sludge. Besides zöogleal bacteria, filamentous forms such as *Sphaerotilus* and *Beggiatoa* are also present, but since they have not been associated with nuisance such as bulking in activated sludge, their occurrence has not been studied in the same detail.

The nature of the bacteria flora will be determined by the nature of the waste. Happold and Key showed that after the application of gas liquor to a bed the bacterial flora underwent considerable modification, bacteria capable of utilizing certain components of gas liquor becoming established. Specific bacteria capable of oxidizing phenols, and thiocyanates, thiosulphates and cyanides have been isolated.

Fourteen well-defined pure organisms were isolated by Harrison from the Monsanto plant treating the trade effluent containing a number of organic compounds; some were found to be specific over the range of compounds tested, whilst others attacked several compounds. One would expect different bacteria associated with different stages of breakdown to be established at different levels of the bed, the heterotrophic forms being nearer the surface and the autotrophic ones nearer the base of the bed.

Although samples of waste taken at different depths suggest that this is so, there appears to have been little work done on the actual distribution of organisms.

Barritt measured the nitrite produced by inoculating sterile solutions containing ammonium salts with effluent from different levels of his sectional bed. He concluded from his results that nitrifying organisms occur in all sections of the bed, but on the basis of the higher nitrite yield by the inoculum from the lower sections, it could be considered that they were more common in the lower portions of the bed.

Using samples of the film removed from different levels in connection with film-accumulation studies, Harkness compared their relative nitrifying capacities and found that although nitrate production was highest in the lower samples, nitrite production was more evenly distributed throughout the different depths, indicating a concentration of nitrate-producing bacteria near the base of the bed, with the nitrite-producers distributed throughout the depth. It should be mentioned that the bed from which the sample was removed was operating on alternating double filtration.

FUNGI

Fungi are of more common occurrence in bacteria beds than in activated sludge. This may be because of the more suitable physical environment or may be because of the constant supply of complex organic matter at or near the surface of the bed; in an efficient activated sludge they would be subjected to starvation conditions for considerable periods.

As saprophytes and primary feeders they are in direct competition with the heterotrophic bacteria and it would appear that the nature and stage of purification of the waste determines which is dominant at the different levels of the bed. With domestic sewage, bacteria usually predominate, but with the introduction of trade wastes, fungi may become dominant especially in the upper part of the bed. Numerous fungi have been isolated from beds but probably the following are the more important ecologically, although other species may be dominant locally under certain conditions: *Fusarium aqueductum, Öospora* (Geotrichum), *Sepedonium* sp., *Ascoidea rubescens, Subbaromyces splendens, Sporotrichum* sp., *Penicillium* sp.

It is of interest to note that the commonly occurring truly aquatic Leptomitus lacteus does not generally appear to have colonized the bacteria-bed environment.

At one works, although it was found in the feed channels in

profuse growths and when detached, blocked the distributor arms, it did not establish itself within the beds. It would appear that the physical environment of the bed determines the species which can inhabit it. Those with tenacious holdfasts such as *Fusarium* and *Geotrichum* are the first to colonize the stones and form a basis for the subsequent establishment of such form, as *Sepedonium* and *Ascoidea*. Painter considered that the different growth-rates and nitrogen requirements of *Fusarium* and *Sepedonium* also contributed to this succession.

Tomlinson found that although *Fusarium* and *Geotrichum* were able to withstand the direct discharge of sewage, *Fusarium* was dominant on the surface and *Geotrichum* below; factors other than the structural modifications were obviously involved. As a result of experiments in which growths were developed in light and dark he found that *Fusarium* was able to compete successfully with the algae *Stigeoclonium* and *Chlorella* in the presence of light, while *Geotrichum* was unable to do so.

In the absence of algae (in the dark) *Geotrichum* competed successfully with *Fusarium* and became the dominant fungus. Thus the dominance of *Fusarium* on the surface of the bed was due to its ability to compete successfully with the algae. He considered that algae might thus limit the growths of fungi on the surface of beds, *Geotrichum* producing thicker mats than *Fusarium*. On the surface of the beds at Birmingham, served by fixed spray jets, a seasonal succession of dominant species has been observed, the bacterial zöogleal growth of the summer giving way to Fusarium which later became overgrown by *Geotrichum* and later, following incipient ponding, by thick growths of *Sepedonium*. The foam-like growths of *Ascoidea* are usually found within the bed.

On beds served with travelling distributors, under the jets of which the sewage impinged on the surface with considerable force, only *Fusarium* and *Geotrichum* became established under jets but as these growths impeded the flow and the sewage spread laterally to the areas between the jets, *Sepedonium* and *Ascoidea* became established in these zones. Factors influencing the amount of fungal growths in beds will be discussed in detail later.

ALGAE

Although algal growths, being restricted to the surface of beds, are not of primary importance in the direct purification processes, their luxuriant growths on the surface of some beds are probably of ecological importance and they may result in the choking of the bed.

The following are the more commonly occurring forms considered ecologically important on bacteria beds:

Cyanophyceae	**Phormidium**
Chlorophyceae	Ulothrix sp.
	Stigeoclonium sp.
	Monostroma.

Associated with these are found unicellular types including diatoms, and, locally, mosses and liveworts may form luxuriant growths.

PROTOZOA

The protozoan fauna of bacteria beds is richer than that of activated sludge. This is probably due to the reduced interspecific competition because of the stratification possible in the beds. Such stratification has been demonstrated by several workers, the species associated with a less-efficient activated sludge being found nearer the surface of the bed and those of a more efficient sludge, nearer the base of the bed, associated with the more purified state of the sewage.

It was also found that the vertical distribution of these species at different works was affected by the strength of the sewage; seasonal changes, due to the sloughing of the film and possibly to temperature changes, also occur. Liebmann found horizontal as well as vertical zoning of microorganisms in bacteria beds.

He also found that the vertical stratification was affected by the loading, the polysaprobic and mesosaprobic forms extending deeper into the bed the higher the load. The common species found in beds are probably the same as those listed for activated sludge. *Carchesium*, which is abundant nearer the bottom of the beds, is of more common occurrence.

FACTORS DETERMINING THE ACCUMULATION OF FILM

The interrelationship and activity of these different members of the film are similar to those outlined for activated sludge, although modified to the extent that stratification is possible. By their combined activity the organic waste is removed or oxidized and as a result the microbial mass increases. Unless means were available for the removal of the excess film the beds would eventually choke.

In discussing the factors influencing film accumulation those affecting both the rate of growth of the film and its reduction must be considered. Within the bed, temperature, food and aeration are probably of primary importance in determining the rate of growth of the film. Increases in temperature up to 20°C result in increased rates of growth for most organisms of the film. Both the nature and strength of the sewage are important in considering the food supply of the film. The nature of the organic waste will determine the dominant organisms of the film.

As has already been mentioned different organisms increase their mass by different amounts whilst oxidizing the same amount of organic matter; zöogleal bacteria, for example, increase less than several common sewage fungi in breaking down a corresponding weight of glucose. The strength of the liquid fed to the bed affects the rate of growth of the film more than does the loading (strength and volume).

Being aerobic organisms, the rate of growth of organisms of the film will be restricted if the aeration of the bed is inadequate, as may occur when film accumulation has taken place or at times when the air currents through the bed are for other reasons reduced.

Agencies operative in the removal of film have been described by different workers. British workers consider that the activity of the grazing fauna is of primary importance in the control of film, although Tomlinson has also reported the significance of bacterial attack on the hyphae of starving fungi.

American workers, on the other hand, consider that physical scouring by the liquid and microbiological activity of the film are most important, the activity of the macrofauna being considered incidental or at most having a minor role.

Usinger and Kellen, however, considered that the larvae of Psychoda were effective in film removal, thereby improving the efficiency of the bed.

It is also possible that with the different films—bacterial, fungal and surface algal growths—and also at different degrees of film accumulation, the relative importance of the different contributory factors varies. Some of the causes suggested by American workers can only be opertative if the film is dominated by fungus and anaerobic decomposition of the film is more likely to be important only when considerable film accumulation has taken place.

A further possible source of misunderstanding is that it has been assumed that the factor causing the seasonal unloading is necessarily the same as that which is responsible for the continuous removal of film which takes place throughout the year.

Observations at Minworth showed that although the grazing activity of *Anisopus fenestralis* was important in controlling fungal film growths in the winter, the seasonal fluctuations in film were not accounted for by the differential effects of temperature on the grazing activity and on the film growth.

The Vertical Distribution of Film

The percentage distribution of film within a bed depends largely upon the volumetric loading and the instantaneous rate of application. Tomlinson found that the film was more evenly distributed in beds operated on alternating double filtration than in single-filtration beds. Increasing the instantaneous rate of application by decreasing the frequency of dosing produced still greater evenness of distribution.

GRAZING FAUNA

Ecologically the holozoic protozoa feeding on other microorganisms should be included in this group, but they are more usually considered as part of the film, as are the small Metazoa such as nematode worms and rotifers. Their importance in controlling the film, however, should not be overlooked because they are less obvious. In experimental beds in which film accumulation was being studied in the absence of grazing fauna, discharge of humus solids was attributed to the large numbers of nematode worms which had accidentally become established.

Of the larger members different species are dominant at different works. Locally other species may dominate: the snail *Physa integra* and other snails are reported from some beds in America in such quantities that they were a serious nuisance by blocking pipelines, by abrasion of sludge-pump pistons and by becoming lodged in the mercury seals of distributors, thereby stopping them. In this country *Lymnaea pereger* is common in the high-rate recirculation beds at Harrogate and its presence there is attributable to the operating conditions.

Lloyd has shown that the grazing fauna of bacteria beds is derived from that of the mud-flats and, these two environments are discussed to explain why so few of the members of the mud-flat community have been able to colonize the bacteria bed. Although toxic trade wastes limit the numbers of species, the chemical nature of sewage is not considered of primary importance in restricting the fauna.

The physical environment and nature of the food are probably the most important factors limiting the macrofauna of bacteria beds. Of the sixteen species of *Psychoda* in the Leeds district, all except one of which are so much alike that they require microscopical examination to identify them, only three are found in bacteria beds.

Their liking for moisture and their ability to breed in confined spaces were considered important factors enabling them to colonize the beds. The unsuitability of the bed for pupation was also considered important in preventing its colonization by many insects. Although described as 'saprophytic' feeders, most grazers should be regarded as holozoic, feeding on the living ilm; they are rarely found amongst the truly saprozoic forms in sewage sludge.

The species which have been able to colonize the bacteria bed find the environment favourable for rapid multiplication. Seasonal temperature fluctuations are less marked in bacteria beds than in other more natural environments and insects which are capable of producing several generations a year do, in fact, produce more generations per year in bacteria beds than under natural conditions.

Because of the natural selection by the environment there are fewer competing species for the available food, but at times when this is limiting severe interspecific and intraspecific competition takes place; under such conditions some—especially the chironomid flies—become predatory.

FACTORS DETERMINING THE NATURE OF THE GRAZING-FAUNA POPULATION

Different beds support different grazing faunas and in determining the nature of the fauna several factors are probably involved. Beds treating normal domestic sewage usually have a full complement of grazing fauna, but a strong sewage tends to limit the fauna; *Psychoda alternata* and *Spaziphora hydromyzina* then become dominant. This may be due to the high oxygen-demand of the sewage or to the nature of the film it produces.

Toxic trade wastes restrict the fauna to few species; different species probably react differently to specific toxic discharges, but generally *Psychoda alternata, Anisopus fenestralis, Achorutes subviaficus, Lumbricillus lineatus* and *Spaniotoma minima* are found in beds treating industrial sewages at Birmingham. At Hudders field Reynold son found the more resistant worm *Enchytraeus albidus* replaced *L. lineatus*. This further restriction in fauna to a few surviving species enables high populations to become established, with the consequent risk of fly nuisance.

Just as physical factors were important in restricting the numbers of species colonizing beds, they are also most important in determining the nature of the fauna in different beds. The nature and size of the medium in the bed has been found to affect the relative composition of the fauna. At one works having beds of different-sized media it was found the larger medium favoured *Anisopus fenestralis* and the smaller medium *Psychoda alternata*; in the very fine medium (½ in.) of one bed

Lumbricus rubellus was dominant. Terry attributed this to a thigmotaxic effect, whereby the organism tends to maintain the maximum contact between its body surface and the surrounding medium. Comparing four different media under identical operating conditions it was found that in the bed having 2 ½ in. round-gravel medium *Lumbricillus lineatus* was the dominant grazer whereas fly larvae were dominant in smaller media.

It is possible that the form of the medium is also of importance: clinker or slag having many pits or depressions would provide a more hospitable niche than smooth gravel, for example. A further physical factor probably associated with the medium size is the downward rate of flow of the sewage over the stones; this obviously varies with the different methods of applying sewage to the beds, e.g. different rates of application and periodicity of dosing. At Minworth (Birmingham), beds receiving sewage as a gentle continuous spray supported a succession of dominant species including *Achorutes subviaticus, Psychoda alternata, Anisopus fenestralis* and *Lumbricillus lineatus*.

Comparison with other beds treating the same sewage, but with increased rates of downward velocity from different operating conditions, showed that as the rate increased, Achorutes subviaticus, Psychoda alternata and Anisopus were successively eliminated leaving lumbricillid worms as the only effective grazing fauna in the alternating double filtration beds on which the distributors revolved once in 30 min, giving a high instantaneous dose and resultant high downward rate of flow.

In beds served by distributors with spaced jets and no splash plates the downward flow varies horizontally across any one bed, being almost dry between the jet-lines in some beds. Although some lateral spread does occur as the sewage passes into the bed, this is not as great as had been thought when the bed is clean, and the upper zone of the bed, where the largest proportion of the grazing population are found, often provides two niches, one below the jet, where the downward flow may be great, and the other where there is hardly any flow.

These conditions produce horizontal stratification of the fauna, lumbricillid worms and some fly larvae being found grazing on the film in the sub-jet zone, and *Achorutes* and other fly larvae (*Psychoda*) being more frequent in the interjet zone, which also provides's suitable conditions for pupation of the larvae from the sub-jet zone. Because of the lack of flow humus tends to accumulate in the inter-jet zone and true saprobic types such as the beetle *Ceryon ustulatus* and *Spaziphora hydromyzina* occur here.

The instantaneous rate of application also affects the vertical distribution of the macrofauna. High instantaneous rates limit some species such as *Psychoda* and *Anisopus* larvae and *Achorutes* in the surface layers, but after the initial flush is spent in the upper layers of the bed, the reduced scouring action permits fly larvae to become established within the bed.

The relative distribution of *Anisopus* larvae in the upper 30 in. of different beds receiving sewage at different instantaneous rates because of different frequencies of dosing. Lumbricillid worms, being strongly prehensile because of their setae, are able to withstand considerable flushing action and their vertical distribution is less affected by the flow, in fact the reduction in fly larvae by high instantaneous rates may result in an increased percentage of worms nearer the surface because of reduced interspecific competition.

Both the effect of medium size and downward rate of flow are affected by the degree of film accumulation. Even under the jets on large round medium where *Psychoda* larvae are unable to exist normally the development of thick fungal growths provides a suitable niche in which they become established. The amount of film accumulation may also determine the type of fauna: *Psychoda,* for example, is usually associated with thick film whilst *Spaniotoma* and *Metriocnemus* prefer a clean bed. It is also possible that the nature of the film, i.e. whether bacterial, algal or fungal, may have a selective effect on the fauna because of food preferences.

Different operational practices may also affect the fauna. Lloyd attributed the absence of *Metriocnemus* in the Barnsley beds to the practice of resting for fortnightly intervals; because *Metriocnemus* is associated with the uppermost part of the bed and requires a constantly wetted surface it was thus adversely affected.

The abundance of any species is largely determined by temperature, food supply and the abundance of competitors and predators. In beds treating domestic sewage, the number of competing species controls the abundance of any one species, but when the fauna is restricted by trade wastes, etc., reduced interspecific competition results in larger populations of the surviving species. Lloyd reported that the abundance of *Psychoda alternata* was successively greater in the beds at Leeds, Barnsley and Huddersfield, the fauna being successively poorer at the three localities.

The seasonal incidence of any species is influenced by the number of other competing species present in the bed; the invasion of the bed by a new species may appreciably change the incidence of the species already present. Reynoldson found that the summer depletion of the fungal film at Huddersfield by *Psychoda* was the chief factor determining the incidence of the worm *Enchytraeus albidus*. The invasion of the Barston beds by *Anisopus fenestralis* was shown to have restricted the incidence of *Psy*choda to August, September and October, whereas previously it was common between April and November.

At higher temperatures, up to the optimum, insects complete their life cycles more rapidly and, other things being equal, the population will tend to increase more rapidly. Because different species have different thermal requirements there is a succession of dominant species throughout the year in beds with a mixed fauna. Lloyd also showed that periods of maximum abundance of sewage flies are not brought about by a gradual increase in numbers, but by a series of alternating peaks and depressions, and the subsequent decline is in like manner; this type of incidence is induced by sudden changes in temperature.

In beds with restricted fauna the food supply in the form of the film is probably the most important limiting factor in controlling populations. At Minworth, where *Anisopus fenestralis* is the dominant grazer for most of the year, its relative seasonal abundance and incidence can be accounted for by the amount of film in the different beds.

It was found experimentally that shortage of food not only resulted in a higher mortality of the larvae due to intraspecific competition, but the larval phase was lengthened and the resultant flies were smaller. Excessive film accumulation, although providing ample food, may create conditions unsuitable for the grazing fauna, the population of which is thereby suppressed.

The value of the grazing fauna in controlling film has already been discussed; it is possible, however, that they play a more direct role in the process. In small-scale experimental beds both Parkinson and Bell and Reynoldson reported increased nitrification following inoculation with *Achorutes* and *Lumbricillus lineatus* respectively. Dyson and Lloyd likened the grazing activity of *Achorutes* to the predatory activity of the holozoic ciliates whose activity maintained the bacteria in an active physiological state.

Their grazing however would be expected to remove bacteria and protozoa indiscriminately as film and according to the "law of disturbed averages" by Volterra if two species are uniformly destroyed in proportion to their abundance, the mean numbers of prey increase and of predators decrease. On this basis the grazing activity of the macrofauna would tend to favour the bacterial population.

The introduction of *Lumbricillus lineatus* to similar experimental beds at Birmingham, which were, however, already clean by virtue of recirculation, produced no increase in nitrification, so it may well be that the increased nitrification reported by previous workers, reflected improved bed conditions. Because insects excrete nitrogen in the form of uric acid or ammonium compounds, readily capable of oxidation, their activity would be expected to assist in the biochemical breakdown of the film.

Unfortunately, although beneficial in the process of purification, the abundance of some members of the grazing fauna gives rise to considerable nuisance on and around sewage works. *Psychoda alternata, Anisopus fenestralis* and less frequently chironomid flies have been reported as causing a nuisance when numerous; although non-biting, because of their close association with sewage, their presence in the home must be considered a potential menace to health.

Culicoides nubeculosus, a troublesome blood-sucking midge, is present on some works. An operator has not only to run his plant to achieve efficient purification, but must do so without causing nuisance in the vicinity; the prevention of fly nuisance is his duty.

FACTORS INFLUENCING THE FILM-GRAZING FAUNA BALANCE

The factors affecting the nature and abundance of the two sections of the bacteria bed community have been discussed. The different populations, however, exist in a dynamic state of balance. The inter-relationship of these populations involved in the breakdown of organic matter in a bacteria bed. The effect of chemical, physical and biotic factors on this balance of populations will now be considered.

CHEMICAL FACTORS

The nature of the organic waste being treated largely determines the nature of the film. This may in turn affect the grazing fauna, which may also be affected directly by the waste especially if toxic. With some wastes, although they are amenable to biological oxidation in bacteria beds, the resultant film is not suitable food for grazing fauna, with a result that the beds eventually become choked. At Stoke-on Trent, experimental beds, successfully treating gas liquor, broke down due to the coating of the medium with a dark-brown resinous matter, probably derived from the higher tar acids present in the liquor.

Apart from the collembolan *Folsomia* sp. a few of which were found near the base of the bed, the bed was devoid of grazers and tests carried out showed that the film was not taken by several common grazers. Because of the differential toxicity to the film and fauna, some wastes, although permitting film growth—usually fungal—suppress the grazing fauna with the result that the bed chokes.

Less toxic substances restrict the fauna to a few species and the reduced interspecific competition may lead to fly nuisance. These chemical factors are imposed by the waste to be treated and apart from pretreatment and trade-effluent control in the case of sewage, the only practical method of varying it is by dilution by such means as recirculation.

Intentional additions of chemicals to the sewage are sometimes practised. When these are made to correct the waste either nutritionally in the case of some trade wastes, or to adjust the pH, they may be considered as modifications to the environment to encourage the suitable organisms for purification. The spasmodic application of chemicals, such as the liming and salting of beds to remove slime and the applications of insecticides in an attempt to alter the biological balance, although necessary palliative measures, are ecologically unsound.

Hawkes showed that, by suppressing the *Anisopus fenestralis* population early in the year by insecticide treatment, the film remained available as food for later generations, which as a result were larger than in the untreated beds where the population was controlled by the limited food supply, the film having been removed by the earlier generation. Although successive applications of insecticide changed the fauna to one dominated by *Achorutes* this could only be maintained by expensive insecticide treatments and when these were suspended the normal fauna was rapidly re-established.

PHYSICAL FACTORS

The size and nature of the medium used in a bed have probably more effect on the grazing fauna than on the film. With large medium excessive film accumulation can be better accommodated, with the result that the efficiency is less affected. However, under clean conditions beds with smaller medium are more efficient. It should be the policy, however, to prevent excessive accumulation of film and not to design to accommodate it, and therefore small medium is the ideal.

Intentional additions of chemicals to the sewage are sometimes practised. When these are made to correct the waste either nutritionally in the case of some trade wastes, or to adjust the pH, they may be considered as modifications to the environment to encourage the suitable organisms for purification. The spasmodic application of chemicals, such as the liming and salting of beds to remove slime and the applications of insecticides in an attempt to alter the biological balance, although necessary palliative measures, are ecologically unsound.

Hawkes showed that, by suppressing the *Anisopus fenestralis* population early in the year by insecticide treatment, the film remained available as food for later generations, which as a result were larger than in the untreated beds where the population was controlled by the limited food supply, the film having been removed by the earlier generation. Although successive applications of insecticide changed the fauna to one dominated by *Achorutes* this could only be maintained by expensive insecticide treatments and when these were suspended the normal fauna was rapidly re-established.

The downward rate of flow of the waste through the bed can be varied at equivalent overall dosage rates by recirculation, double filtration, the frequency at which doses of the waste are applied to the bed, and the type of nozzle through which it is discharged. Although this factor affects both film and fauna, the effect on the film is largely restricted to the surface and below the surface the fauna is affected to a greater extent than the film.

It has been claimed that film accumulation can be prevented by the flushing action of the sewage and the success of alternating double filtration and low-frequency dosing, i.e. heavy doses at infrequent intervals, has been attributed to this physical scouring because of the high instantaneous dosage rates. Although these forces may be operative in removing humus solids previously detached by other means such as grazing, fungal growths are able to withstand very great scouring action as is shown by their presence on the impeller blades of rotary pumps revolving at approximately 1000 r.p.m.

Although immediately below the jet the fungal growths may be limited to *Fusarium* and *Geotrichum*, below the surface layer the downward flow of the liquid, although in some cases great enough to affect the grazing fauna, is not considered to have any significant effect on the film within the bed. Thus, because of the differential effect of the rate of downward flow, the film accumulation may be greater at higher instantaneous rates of dosing.

Testing six types of distributor nozzle through which the sewage was discharged to the bed at different forces and differently distributed it was found that because of the suppression of the grazing fauna where the flow was great the film accumulation was greatest. Where horizontal stratification occurred due to jet spacing the two niches thus provided permitted *Anisopus* larvae and *Achorutes* to exist together. Where the distribution was even, these two grazers were in direct competition, Achorutes only becoming abundant when the *A. fenestralis* population was reduced in the summer. The continued grazing activity of the two populations is considered desirable and on this basis, jets spaced at intervals, were considered preferable to even distribution. Temperature also has a complex differential effect on the film fauna balance. The rate of increase and activity of both film and fauna is generally greater at higher temperatures within the range usually experienced in beds. Therefore the higher general metabolic activity within the bed results in enhanced purification at higher temperatures. Low temperatures suppress the grazing activity more severely than the growth of film and this differential effect has been claimed to account for the winter accumulation of film. With a purely bacterial film this is probably the case.

The relationship between fungal film and grazing fauna; although successive generations of fly larvae modulated the film incidence, they were not the cause of it. Whatever the cause, however, the winter accumulation of film and reduced efficiency are associated with lower temperatures and in many cases these are a limiting factor in the winter operation of beds. Siting of the beds and the temperature of the waste both affect the bed temperature.

If the aeration of a bed is brought about by the differences in temperature throughout and above the bed, then because of the greater seasonal fluctuations in air temperature than in bed temperature, the aeration will vary throughout the year, usually being greater in winter than summer. This increased ventilation may result in increased fungal growths in the winter which then reduce the aeration.

BIOTIC FACTORS

The most important biotic factor affecting the balance of populations is the basic food supply entering the community, i.e. the waste. The strength of the waste determines the growth rate of the film which in turn determines the grazing fauna population. The growth rate can, however, be controlled by controlling the feeding as opposed to the food itself.

Although the film population and fauna population mutually affect each other, it is probably true to say that the amount of film is a more important factor determining the fauna population than the converse. In more natural environments, factors controlling animal populations are still not fully understood, but in bacteria beds the food supply is the most important factor. The soundest way to control a fly population breeding in the bacteria beds, therefore, is by controlling their food supply, i.e. limiting the accumulation of film.

The biological control of insects by the introduction of competitors or predators, operating as it does on a density-dependent factor, appears attractive for application in the enclosed environment of a bacteria bed. Although this method met with initial success in other fields, especially when applied to confined environments with an equable climate, such as tropical islands, it would appear that its applications are now becoming exhausted.

The fly *Spaziphora hydromyzina* is predatory on *Anisopus fenestralis* but, presumably because of its temperature requirements, it occurs later in the summer than the peak *Anisopus* population; it in turn is also highly parasitized by an ichneumon fly *Phygadeuon cylindraceus*.

ACTIVITY IN BACTERIA BEDS

As with activated sludge, a pre-requisite of the process is the intimate contact of the waste with an active microbial population in the presence of oxygen. In bacteria beds, however, this is achieved in a different manner by allowing the waste to flow over the micro-organisms exposed as a zöogleal film on the bed medium, the oxygen mostly being provided by the air currents through the bed. The biophysical and biochemical processes bringing about the waste removal and its oxidation are probably similar to those operative in activated sludge.

Bacteria-bed film has been credited with phenomenally high instantaneous rates of oxidation in keeping with the rapid removal of B.O.D. during the relatively short retention time of the liquor in the bed. The relative instantaneous rates of removal and oxidation will probably depend upon the nature of the waste. With wastes having the organic matter in solution and in such concentrations that they are limiting the growth rate of the micro-organisms, the organic matter could be expected to be utilized as it is removed.

With stronger soluble wastes some may be stored as reserves such as glycogen within the cells for subsequent oxidation. With wastes having organic matter as suspended or colloidal solids, these readily flocculate on the surface of the film and may only be absorbed slowly, prior to oxidation.

The period of contact, the liquid retention time, should be sufficient to ensure the transfer of the waste to the film by adsorption or absorption for subsequent oxidation. To avoid the progressive accumulation of unoxidized waste in the bed, the overall rate of oxidation of the waste must equal the rate of removal of the waste.

In other words the amount of waste, the oxidation of which is completed per day, should equal the amount of waste removed per day. This does not imply that the time needed for oxidation is the same as that required for the initial removal of the waste. There is no functional relationship between the liquid contact time' the time required for the removal of the waste—and the instantaneous rate of oxidation of the removed waste. As an analogy we may consider the routine in a laboratory regularly receiving 100 samples for B.O.D. determinations each day.

Although the time needed to perform the test on 100 samples is 5 days, the overall rate of obtaining results is the same as that of the reception of samples—100 per day. The 100 samples received per day represent the waste removed in unit time in the bed, the 5 day incubation period of the test represents, conveniently enough, the period between removal of the waste and the completion of its oxidation. The 100 results represent the amount of waste, the oxidation of which is completed within unit of time. The 500 samples in the incubator represent the amount of waste stored or under oxidation in the bed.

The analogy may be safely carried a step further. If the number of samples were increased to 120 per day, then, after a 5 day period, the number of results—or tests completed, would be 120 per day provided that two conditions were fulfilled. Firstly the capacity of the incubator must be adequate to accommodate the increased number of samples being incubated—600, and secondly, the laboratory staff, by increasing its activity or numbers, must be capable of dealing with the extra samples.

Similarly, in a bacteria bed a 20 per cent increase in the load of removed waste could be dealt with provided that the film was able to accommodate the resultant increased storage matter without impairing its efficiency and secondly the micro-organisms were able to increase their activity or population to deal with the increased food supply. These considerations point to two factors limiting the loading of bacteria beds; the extent to which they can accommodate the unoxidized removed waste without impairing the efficiency of the film, and the size of the microbial population that the bed can effectively support. In addition, of course, the effect of the increased load on the initial removal must also be considered.

The instantaneous rate of oxidation is a function of the concentration of the substrate, thus the high rates of oxidation associated with the bacteria-bed film may be due to the concentration of the waste at the surface of the film due to the Physical forces associated with interfaces. The amount of unoxidized organic matter in a bed will depend upon the speed of oxidation.

If the speed is halved the amount of unoxidized matter is doubled but providing that the film can accommodate this increase and the microbial population deal with it, the total amount oxidized per day will be the same, for although the instantaneous rate of oxidation is halved the amount available for oxidation is doubled.

Thus a new balance is established. Below 50°F although the initial removal rate in bacteria beds is probably not affected, the instantaneous rate of oxidation probably is. In beds treating wastes having a high proportion of suspended or colloidal matter, this may result in an excessive accumulation of unoxidized flocculated solids in the film. This resultant increase in film accumulation at low temperatures should be distinguished from that due to the increase in microbial mass, for reasons already discussed.

As with activated sludge it is necessary to maintain an active population of heterotrophic micro-organisms and in a bacteria bed this can only be achieved by preventing excessive overgrowth of the film. There is normally a seasonal decline in the efficiency of beds during the winter, in contrast with the activated sludge process.

We have seen that there is usually more film present in the beds during the winter than in the summer. The extent to which the decreased efficiency is due to lower temperatures or increased film is difficult to assess. The seasonal fluctuations in film accumulation, temperature and efficiency of two beds, one in which the usual accumulation of film occurred in the winter and the other in which it was suppressed by controlled frequency of dosing.

It will be seen that the efficiency of the bed in which there was no winter accumulation of film showed little seasonal fluctuation compared with that in the bed in which winter film accumulation occurred. Although not eliminating temperature as a factor, these results do confirm that film accumulation itself is an important factor.

Whilst most workers agree on the desirability of preventing excessive accumulations of film, there would appear to be some difference of opinion as to the optimum amount needed. Some consider that as thick a growth as possible, compatible with the prevention of ponding and ensuring the adequate aeration of the bed, is desirable.

Lloyd considered that were it not for the increased metabolic activity of the micro-organisms at higher temperatures in the summer, their reduced numbers, due to the dominance of the grazing fauna, could result in decreased purification. Experience with beds at different works, however, has led me to believe that a very thin film indeed, perceptible only to the touch as a slime on the stones, is all that is necessary for efficient purification. This view is supported by experimental evidence.

Using inclined rotating tubes it was found that the maximum

efficiency of removal of organic carbon was reached when a film of only 0.12 mm average thickness was formed. Wuhrmann calculated that the thickness of film in which all the bacteria could be expected to receive adequate oxygen supply, as being 100-200 $_{U}$+03BC (0.1-0.2 mm). Thus although the amount of film which can be accommodated in a unit volume of bed, without causing ponding or impeding aeration, is limited by the void capacity between the media, the amount of active film is limited by the total surface area of the stones on which the thin film can be supported.

Although only a thin bacterial film is considered necessary for efficient purification, the thickness at which its activity per unit area is impaired is more difficult to establish in practice. As a rough guide, considering medium of nominal size, the volatile solids should probably not exceed 50g dry weight per cubic foot of bed, possibly less if nitrification is desired. When the growth is dominated by certain fungi, such as *Ascoidea* which forms foam-like masses filling the interstices, larger amounts—up to 200g dry weight volatile matter per cubic foot—may be maintained in a healthy aerobic condition.

Although such growths render the bed highly efficient, unless they are controlled they eventually result in ponding with marked decrease in efficiency. Such growths have been experienced within beds operating on A.D.F. Other fungi such as *Fusarium* and *Geotrichum* are, like bacteria, more closely associated with the surface of the stones.

Consideration must also be given to the vertical distribution throughout the depth of the bed. In conventional beds, because of the reduction in concentration of the waste during purification as it passes downwards, most growth occurs at or near the surface. Although at times when such surface growths are sufficient to reduce the efficiency of the affected uppermost part of the bed, the film within the bed may still be in the active log phase.

The surface growths, however, may cause ponding which interferes with the distribution of the waste and the aeration of the bed, thus affecting the efficiency of the whole bed. Furthermore, because of the reduction in efficiency of the upper portion of the bed, the film below receives a higher concentration of the waste, thereby increasing the growth rate of the organisms. Thus a downward growth of the film takes place, the efficient portion of the bed being successively reduced and the nitrification zone lowered or even eliminated. During the unloading of the film the reverse probably occurs, the lower accumulation of solids being discharged first.

Bacteria beds must not only be operated efficiently but also without causing nuisance. Flies, such as *Psychoda* spp. and *Anisopus fenestralis*, have caused serious nuisance in the vicinity of bacteria beds. Synthetic insecticides are being used successfully to control these flies but such a practice, although under many conditions necessary as a palliative measure, is ecologically unsound.

The occasional application of insecticide results in a reduction in the insect population causing an upset in the ecological balance which may result in a rapid increase in film which is then available to support an increased fly population necessitating further applications of insecticide. It is possible by successive applications of insecticide to change the fauna of a bed from one dominated by flies to one having less troublesome organisms such as *Achorutes* or *Lumbricillid* worms as the dominant members.

Such a balance can, however, only be maintained by expensive insecticide treatments and when these are suspended the normal fauna is rapidly re-established. A further objection to the use of insecticides on bacteria beds is the danger of discharging a toxic effluent.

In more natural environments the factors controlling animal populations are still not fully understood, but in bacteria beds food supply is a most important factor. The soundest way to control a fly population breeding in bacteria beds, therefore, is by controlling their food supply, i.e. limiting the accumulation of film. Thus, for optimum efficiency and prevention of nuisance, the limitation of film accumulation is desirable.

The control of the microbial population, necessary for efficient operation and prevention of nuisance, cannot be directly effected by such mechanical operations as adjusting the excess sludge weir as in the activated sludge plant. It can only be achieved indirectly through biological agencies by the ecological control of the competing populations.

PRACTICAL METHODS OF MAINTAINING THE BALANCE OF LOW LEVEL

The rate of film accumulation may be considered as a function of the rate of growth and rate of removal.

$(R_a) = (R_g) - (R_r)$ *Rate of accumulation Rate of growth Rate of removal*

To maintain R_a at zero, i.e. to prevent accumulation, two theoretical alternatives are possible. One method would be to operate the bed to encourage maximum growth—i.e. R_g is a maximum—for the waste being treated, and maintain conditions so that $R_r = R_g$. The alternative would be to limit the growth rate R_g by operational means so that R_g equals Rr under the operating conditions imposed.

The two alternatives are thus basically different and are, as we shall see, in some ways incompatible. Of the two possible approaches, the former, whereby maximum growth rate and activity of the film are encouraged, but the accumulation of film is maintained at a low level by a dominant grazing fauna, is preferable. The resultant low level of the film accumulation would cause severe competition for the limited food among the grazing fauna.

In insect populations the percentage number of larvae which successfully pupate, before emerging as adults, decreases with decreasing food supply when the latter is limiting. Thus under conditions of limiting food supply, much of the available food is wasted to the fly population by the death at different stages of development of larvae which have used some of the available food. This phenomenon, known ecologically as a "scramble", is attractive as a method of utilizing the benefits of an insect population without incurring the nuisance of a large adult population.

It is a self-regulating mechanism in that the fewer eggs produced by the reduced adult fly population would result in a reduced larval population with less severe competition and a resultant lower death rate. Unfortunately, it is only in beds treating a weak waste with resultant low growth rate —(R_g) that conditions suitable to support this balance can be maintained throughout the year.

In other beds, although at summer temperatures such conditions may prevail, the lower winter temperatures result in the accumulation of film. If the rate of accumulation is slight, it may be that the resultant accumulation throughout the winter is not sufficient to affect the efficiency of the bed before the spring unloading takes place.

In more heavily loaded beds, or those subjected to lower temperatures, the greater accumulation of film results in decreased efficiency. In both, the unloading in the spring, usually brought about by an increased grazing population, uncontrolled by the available food, is associated with a potential fly nuisance. Also, by permitting this winter accumulation of film, the increased rate of discharge of humus sludge during the resultant spring unloading, imposes an increased load on the digestion plant, often when it is least able to cope with it.

These effects of low winter temperatures could be overcome by heating the beds either directly or by heating the feed liquor. Fuel for this could be provided by the gas from the digestion of sludge. Alternatively in the vicinity of generating stations, cooling water could be used.

In many beds, because of the nature and strength of the waste and the economic necessity of operating at higher rates, the grazing fauna is not able to dominate the film, especially in the winter, and in such cases it is necessary to adopt the second alternative and control the growth rate (R_g). Of the advantages claimed for the modified methods of filtration commonly practised, such as recirculation, alternating double filtration, and low frequency dosing, the control of film accumulation is common to all.

The method by which this is achieved, however, has been variously described; some consider that the flushing action resulting from the higher rate of instantaneous application is the cause, others consider that nutritional control is the operative factor. In the light of results reported above, it is considered that flushing action can have little effect in removing attached film from within the bed at the usual rates of application practised in this country.

An understanding of the controlling mechanism is not only of academic interest but is essential for its successful application and for the diagnosis of the cause of failure should this occur. We shall, therefore, now consider these different methods of "filtration" as film controlling processes, although in doing so their other advantages should not be overlooked.

Recirculation

Ecologically the most important features of recirculation, whereby the waste is diluted with returned effluent before being applied to the bed, are the reduction in strength of feed and the increase in the hydraulic load. These two factors influence the growth rate of the film and its vertical distribution throughout the depth of the bed. Applying the growth rate/concentration curve discussed earlier, a waste having a concentration C_1 gives an arbitrary growth rate R_1 on the surface of the bed operating as a single filter.

By introducing recirculation and thereby reducing the concentration of the feed to C_2 the resultant growth rate is reduced to R_2. Consider now the growth rate within the bed, say at a depth of 2 ft. With single filtration suppose the initial waste concentration C_1 has been reduced to C_3 giving a reduced growth rate R_3. With recirculation because of the higher hydraulic loading the reduction in concentration in the upper 2 ft of bed is less than with single filtration and the resultant concentration C_4 may well be higher than the concentration C_3 at the corresponding depth in the bed operating by single filtration, as indeed was found to be the case by Lumb and Eastwood.

As a result, the growth rate of the film at a depth of 2 ft within a recirculating bed is greater than in a conventional bed. These results would account for the reduction in amounts of surface film and the more even vertical distribution of the film throughout the depth of the bed with recirculation. A greater depth of the bed is thus used for carbonaceous oxidation but the zone of nitrification is restricted.

To effect this reduction in film growth at the surface it is essential that the dilution of the feed is sufficient to result in a concentration which falls on the slope of the growth rate curve, i.e. a concentration at which nutrient is limiting. It could be that with a strong waste C_5, and a low dilution ratio, the resultant diluted feed C_6 would have the same growth rate—$R_{5/6}$ at the surface of the bed, the concentration of the feed C_5 not being growth controlling. Furthermore, because of the resultant greater hydraulic loading it is possible that the accumulation of film will also occur to a greater depth than with single filtration.

To control film growth nutritionally in beds treating fluctuating flows, the dilution ratio should ideally be constant, i.e. the recirculation rate varied directly with the flow of waste. Lumb and Eastwood compared different applications of recirculation to beds treating fluctuating sewage flows. On the bed in which the recirculation rate was varied directly with the sewage flow to maintain a constant dilution ratio, there was little ponding. On a second bed, however, in which the recirculation rate varied inversely with the sewage flow, to maintain a constant flow to the bed, severe ponding occurred in the winter period.

The operation of the former ideal method of recirculation involves plant of larger capacity and variable pumping capacity with the necessary control mechanism. A third method investigated by Lumb and Eastwood, whereby a constant recirculation rate calculated to give an average dilution ratio equal to the ideal system was maintained irrespective of the sewage flow, proved only slightly inferior to the ideal. Plant capacity is not as great as in the ideal system and pumping plant and control gear is the simplest of the three methods. The degree of dilution of the feed depends not only upon the dilution ratio but also on the quality of the diluting effluent.

Thus, if for any reason the efficiency of the bed falls off, normal recirculation will be less effective in controlling film, and further deterioration in conditions may result. By diluting the feed with purified effluent from a more efficient bed or activated sludge plant it is possible that further deterioration would be prevented. Besides this effect of the growth rate (Rg) recirculation may also affect the removal rate (R_r) by changing the numbers and species of grazing fauna.

The increased hydraulic rate of application tends to reduce the numbers of fly larvae such as *Psychoda* and the Springtails such as *Achorutes*. The more prehensile *Lumbricillid* worms persist and with the reduced interspecific competition they may increase in numbers.

Such a change in fauna has been reported by Lumb and Eastwood and by Hawkes. There were indications that the *Lumbricillid* worms were less active in removing film than the *Psychoda* population they had replaced.

ALTERNATING DOUBLE FILTRATION

The process of A.D.F., in which waste is treated on a pair of beds in series, each bed alternately becoming primary and secondary in successive periods, was first conceived by O'Shaughnessy as a means of controlling excessive film growth. He had observed that excessive accumulations on beds could be removed by applying the effluent from a bioflocculation plant.

Thus, from the beginning it was acknowledged that the control of film was an important feature of A. D.F. Tomlinson later demonstrated that this was due to a starvation effect during the secondary stage of the process during which the film decreased in amount. This is also explicable in terms of the growth curve. During the primary stage the growth rate (R_1) at the surface is determined by the concentration C_1 of the waste, and is the same as that on the surface of a conventional bed treating the same waste. During the secondary stage the film is subjected to the settled primary bed effluent of concentration C_2 giving a negative growth rate R_2. Thus the resultant growth rate

$$\frac{R_1 + R_2}{2}$$

is less than R_1, the rate at the surface of a conventional bed. Because of the higher hydraulic loading the film is distributed more evenly throughout the depth of the bed as in recirculation. The total amount of film, however, is less than in single beds treating the same waste since with A.D.F. all the film is subjected to the negative growth phase in the secondary stage, whereas in a conventional bed the bulk of the film in the upper layers of the bed is normally never subjected to starvation conditions and the resultant negative growth rate.

Thus, this principle of A.D.F. is based on a fundamentally sound theoretical basis. The process has been successfully applied to the treatment of sewage at Reading and Derby and has proved of particular value in treating wastes, such as those from milk processing, which tend to give rise to fungal growths.

Experience at Birmingham, where the successful development of the process by the Water Pollution Research Laboratory led to it being adopted for large scale operation, has shown that to be successful the process must produce a primary effluent which will ensure that in the secondary stage the film is in the endogenous phase to produce a negative growth rate (R_2).

Even when this is achieved the resultant growth rate is positive and to prevent film accumulation this has to be countered by an equal removal rate (R_r). In beds operating at conventional hydraulic loadings this is mostly provided by the activity of the grazing fauna. At times when this is reduced, as when temperatures are low, it is essential that the resultant growth rate is kept low by a negative growth rate in the secondary stage.

If for any reason an inferior primary effluent is produced either because of reduced efficiency or higher loadings, the carbonaceous oxidation is continued in the secondary stage and the final effluent is at first little affected apart from the reduced nitrate content since nitrification is not similarly increased in the secondary stage.

However, the higher nutrient concentration results in a positive growth rate in the secondary stage and an increase in the degree of film accumulation. This could result in a further decrease in efficiency and a further deterioration in the primary effluent. Thus conditions progressively deteriorate until the process eventually breaks down in the complete ponding of the beds. It was found that this adverse sequence of events could be prevented by applying a further growth-controlling factor—the frequency at which successive doses of sewage were applied to the bed.

FREQUENCY OF DOSING

At the same overall rate of treatment the waste may be applied as a continuous spray or as a succession of doses. The frequency at which these doses are applied may also vary, the lower the frequency the higher the instantaneous rate of application. This frequency of dosing is a most important ecological factor in the operation of bacteria beds. Early in the history of bacteria beds it was reported in evidence to the Royal Commission on Sewage Disposal that intermittent dosing of beds limited the accumulation of film.

It is of interest to note that one of the first bacteria beds to be operated at the Lawrence Experiment Station, Massachusetts, in 1890 was dosed intermittently at 20 min intervals. At Birmingham early in the century, one of the experimental beds was equipped with a one-armed distributor mechanically driven to rotate once every 7 min by means of a 2 h.p. oil engine.

Lumb and Barnes at Halifax found that by slowing down the speed of rotation of a four-armed distributor, the condition and performance of the bed was improved. Tomlinson and Hall investigated the effect of frequency of dosing on beds operating on A.D.F. They concluded that for four-armed distributors the optimum rate of rotation under their conditions of experiment was between 15 and 30 min per rev.

During the later period of these investigations comparisons of the seasonal fluctuations in film accumulation and fauna in low and high frequency dosed beds were made. In the beds served with rapidly revolving self-propelled distributors there was, each winter, an excessive accumulation of film; in similar beds treating the same sewage at the same rate but having the distributors mechanically rotated at speeds between 30 and 55 min per rev, the amount of film was uniformly low.

There was also a marked effect on the grazing fauna; the fly populations, both *Anisopus* and *Psychoda,* were almost eliminated in the low frequency dosed beds. This was partly due to the reduced food supply in the form of film and also to the greater downward rate of flow of the sewage suppressing the larvae populations within the bed.

The more strongly prehensile *Lumbricillid* worms were better able to withstand the stronger flow of sewage and, in the absence of competition from the fly larvae, they became more abundant in the low frequency dosed beds. Thus, by reducing the frequency of dosing, ecological conditions were created which not only maintained a low degree of film accumulation necessary for efficient purification, but also prevented fly nuisance by changing the fauna from one dominated by flies to one dominated by worms.

For successful application of controlled frequency dosing it is essential to appreciate the principles underlying the process. Unfortunately, as in many developments of biological filtration where practice has led theory, the practical success has been explained by various theories.

The resultant high instantaneous dosage rate is generally considered to be an important feature of the process. In view of the evidence reported earlier in this work, however, the scouring effect of this high instantaneous rate cannot be considered to be an important factor in controlling film, especially within the bed, although it does affect the fauna, the fly larvae being replaced by the more strongly prehensile *Lumbricillid* worms. An alternative theory suggests that because of the infrequent application of waste, the film is subjected to reduced nutrient concentration, if not starvation conditions, for some of the period between doses.

To investigate this nutritional effect, samples were taken at minute intervals from side arm sampling channels collecting the effluents at 1 ft, 3 ft and 5 ft depths of the bed to measure the change in concentration between closes at 15 min intervals as measured by the O.A. Similar samples from the same bed being dosed at 1 ½ min intervals were taken on the next day. The results of both tests expressed as ppm O.A. remaining per 100 ppm.

Only slight fluctuations in the strength of liquid occurred at any one depth with the rapidly revolving distributor, when the distributor revolved slowly there was a marked reduction in strength over the 15 min period between the doses, this being most marked at the 1 ft depth. Applying these results to the conventional growth rate curve it is seen that at the 1 ft level, for more than two-thirds of the time the growth rate with low frequency dosing is considerably less than with high frequency dosing and may well pass into the endogenous phase towards the end of the period between doses.

Within the bed at a depth of 3 ft for half the time the growth rate would be higher with high frequency dosing, although at 5 ft the growth rate would be higher with low frequency dosing. This would account for the more even distribution of film throughout the depth of the bed with low frequency dosing.

The results illustrate that with low frequency dosing the average concentration of nutrients available to the film in the upper layers is reduced, thus reducing the growth rate. Therefore the control of film in low frequency dosed beds could be accounted for by nutritional control as in recirculation and A.D.F.

Suggestions by Stanbridge that low frequency dosing may have advantages other than those attributable to the control of film have been investigated. To investigate this possibility, use was made of a circular bed which had a distributor fitted with variable speed drive. On several daily occasions the distributor was revolved at different speeds between 1 rev in 3 min and 1 rev in 60 min, the speed on each occasion being chosen at random; the efficiency, as assessed by the percentage O.A. and B.O.D. removal and the amount of oxidized N in the effluents, was determined. On the same occasion the efficiency of a similar bed receiving the same sewage through distributors revolving at a fixed rate was measured.

The differences between the efficiency of the two beds on each occasion plotted against the rate of revolution of the distributor demonstrates any effect of frequency of dosing, apart from those resulting from film control, since during the short period of the tests the amount of film could be regarded as constant. There was, as expected, over the greater part of the range, between 15 min and 60 min, a reduction in efficiency resulting from decreasing the frequency of dosing.

There was, however, evidence that the optimum rate of revolution, for a clean bed, was between 10 and 15 min for a four-armed distributor with staggered jets. At speeds of revolution greater than these, there appeared to be a lowering of efficiency. It is of interest to note that nitrification was only slightly reduced by low frequency dosing. Results quoted by Levine and Byrom also showed that the efficiency as measured by the percentage B.O.D. and O. A. removal respectively, was somewhat higher when dosed at 5-7 min and 8 min intervals respectively than at 2-2 min.

It does not follow from the results that the optimum speed of rotation was 10-15 min per rev; to control the film at limits which do not adversely effect the efficiency, it may be necessary to revolve the distributor at speeds slower than once every 15 min, the resultant loss in efficiency more than being made good by the gain in efficiency resulting from the prevention of the film accumulation. The results nevertheless do indicate some advantages other than those attributable to film control, at least over the range 1-15 min per rev.

Some such possible benefits of low frequency dosing have been discussed by Stanbridge. He suggested that although the time of retention of some of the liquid is less with low frequency dosing this may be more than offset by the longer retention time of the liquid held interstitially between doses. Work reported by the Water Pollution Research Laboratory has shown that the mean retention time may be increased by decreasing the frequency of dosing.

Below a certain frequency of dosing, however, the retention time is, of course, reduced. Liquid retention time itself may not be so important providing it is sufficiently long to enable the waste to be transferred to the film. The oxidation of the removed waste can proceed during the period between doses.

In attempting to determine the times of contact at different rates of revolution it was observed that within a minute of the passage of the distributor arm over the section of the bed a flush of effluent was discharged from the bottom of the bed. The peak concentration of the chloride appeared in the effluent within the first minute but was still being discharged in decreasing amounts three hours later.

The reduction in chloride concentration in the effluent took place not at a steady rate but in a series of regularly timed steps; a drop in concentration coincided with the surges of effluent which immediately followed the passage of a distributor arm over the surface. It would appear that when sewage is applied to beds intermittently it passes into the bed as a surge mixing with the interstitial liquid; some of the mixed liquid is displaced in the form of a flush of effluent; the remainder is held interstitially until the next dose, only a small quantity draining away during this period.

The amount held interstitially within the bed is, therefore, important in determining the efficiency of a bed intermittently fed, because whilst the liquid was held in the bed, purification would be proceeding. It may well be that such a bed would be more efficient if constructed of medium smaller than the conventional medium used today; the low film concentration associated with the process would make this possible.

Low frequency dosed beds are thus quite different hydraulically and in the course of purification from conventional beds. In the latter the waste is considered to be oxidized as it percolates downward over the medium, this course of oxidation at different depths being expressed mathematically by Velz's equation. In low frequency dosed beds the course of purification must be considered more in relation to time than depth. Certainly the concept implicit in the terms "percolating" or "trickling" filters is not applicable to low frequency dosed beds, which are better considered as surge filled aerobic contact beds.

In view of the rapid mixing of feed and interstitial liquid they are in this respect analogous with the modern complete mixing activated sludge processes. Danckwerts, comparing the piston and complete mixing types of continuous flows through processing units generally, considered that because of the greater spread of the residence time with complete mixing and assuming the rate of the reaction falls off with the extent of the reaction, complete mixing types of reactor were less efficient than the piston flow type.

With autocatalytic reactions (such as biological oxidation), however, it was considered that, because of the different rate-of-reaction curve, the completely mixed reactor may be superior. In applying these views to bacteria beds and activated sludge plants, however, it should be borne in mind that the lag phase is probably absent.

A further advantage of low frequency dosing, it was suggested, is that the higher instantaneous rate of dosage results in a greater volume of the bed medium being wetted and thereby the effective capacity of the bed is increased. To investigate this suggestion, the lateral spread of sewage applied to the surface of a clean bed as jets 12 in. apart was assessed by collecting and measuring the sewage in twelve 2 in. wide adjacent trays placed parallel (tangential) to the travel of the jets so as to collect the liquid from an area 1 ft × 2 ft.

Measurements were made immediately above the surface of the

bed and at three depths within the bed and at different speeds of revolution of the distributor arm. Most of the lateral spread occurred as the jet impinged on the bed surface; there was, however, no indication that with slower speeds of revolution the lateral spread was greater. Although other advantageous effects of low frequency dosing may exist, it is considered that the control of excessive film accumulation, with the resultant increased efficiency, is probably the most important effect.

An appreciation of the nutritional method of film control operative in low frequency dosing and the resultant beneficial effects on bed efficiency, makes possible the reasonable application of the principle. No increase in efficiency could be expected of a bed in which film accumulation never occurred, by slowing down the rate of revolution of the distributor to speeds slower than 10 min per rev for example; slower rates of revolution would probably result in decreased efficiency due to decreased retention times.

For a similar reason, because film accumulation is usually associated with winter conditions, the frequency of dosing necessary to suppress winter accumulations could be lower than the optimum required for summer conditions. On this basis, variable frequency of dosing is desirable. Whether the extra cost of providing variable speed drives is justifiable will depend upon the degree of deterioration in efficiency during the summer due to the low frequency dosing.

At Ewell, for example, Stanbridge considered that the decrease in efficiency in the summer months which resulted from low frequency dosing was not sufficient to justify changing the speed of rotation, which during the winter gave marked improvements in efficiency.

A bed already ponding cannot be immediately cured by merely slowing down the speed of rotation of the distributor; the periods of low waste concentration between doses, necessary for nutrient control, would not occur under conditions of continuous ponding. If the success of low frequency dosing is due to nutrient control, then it can only be effective in controlling a film of living organisms, fungal or bacterial, and not against an accumulation of non-living solids. The possible accumulation of flocculated solids in beds treating wastes with high suspended and colloidal matter during the colder periods when oxidation is retarded.

The overall effect of low frequency dosing on such beds is not fully known. One could expect that apart from nutritional control of the organisms, the resultant higher instantaneous rate of application would result in a deeper penetration of the flocculation zone to within the bed where temperature would be higher.

It is also possible that the higher instantaneous rates of application could flush away some of the flocculated solids even though the living film resists such action. Similarly, if solids applied with the waste or those formed chemically or physically within the beds accumulate, they cannot be controlled by nutritional methods and low frequency dosing could only succeed by virtue of the high instantaneous rates of application as in the case of solids flocculating within the bed.

Practical Methods of Controlling the Frequency of Dosing

In determining the frequency of dosing—best expressed as the time in minutes between successive doses—consideration should be given to unit small areas representative of one stone on which the microorganisms live. Thus on a circular bed served by a four armed distributor, the jets of which are not less than 1 ft apart and are staggered on all four arms, the frequency of dosing at the surface is the time of one complete revolution of the distributor.

Even at a depth of 1 ft the degree of lateral spread results in the frequency of dosing at this depth, being half the time of revolution of the distributor. It is only within the depths of the bed where the lateral spread of the waste ensures that the flow from one jet is evenly distributed over at least 1 ft that the frequency of dosing can be considered as one quarter of the time of revolution. Troublesome growths usually occur in the upper layers of the bed and in this respect circular beds having distributors with staggered jets are capable of creating conditions of low frequency dosing in the upper layers, to prevent film accumulation and a higher dosing frequency within the bed.

Rectangular beds, having the conventional reciprocating distributors, have the same frequency of dosing at all depths. Also, when dosing in both directions, the frequency of dosing differs at different positions along the length of travel. By having distributors with two arms, the jets in which are staggered, and which dose alternately in opposite directions, the conditions of the circular beds could be approached. The jet spacing should be sufficient to ensure that in the upper layers of the bed each unit small area receives waste mostly from one arm only.

This can be ensured by having half the width of a bed dosed in one direction and the other half dosed in the opposite direction but the frequency of dosing is then the same at all depths. Low frequency dosing can be applied to conventional single filtration or to supplement the growth controlling effects of recirculation or A.D.F.

The optimum frequency of dosing will depend upon the different operating conditions and the growth promoting properties of the waste. The optimum frequency may vary seasonally, thus making some variability in dosing frequency desirable. In such two stage processes as A.D.F. It is possible that the optimum frequency of dosing of the primary and secondary stages differs; a lower frequency being necessary to control film growth in the primary stage.

The frequent alteration would be made practicable by providing variable frequency of dosing. Generally by providing for a dosing frequency range of 10 to 30 min in the upper layers of the bed, most conditions of operation should be covered. The conventional reaction-jet propelled distributors on circular beds rotate at rates varying with the flow, but usually faster than 1 rev in 3 min. This rate of rotation can be decreased somewhat by turning some of the jets to discharge in the opposite direction to the driving jets and thus act as a brake.

The extent to which this method can be carried is limited by the need to ensure rotation of the distributor at low flows and under windy conditions. Generally it is not possible to satisfactorily reduce the speed of rotation to rates slower than 1 rev in 10 min by this method or by other methods involving a non-differential braking device.

Rotating distributors, propelled by a water-powered wheel travelling around a circumferential track or rail, usually have a slower rate of rotation and are more amenable to some adaptation for lower frequency dosing, by changing the travelling wheel for one of smaller diameter, for example. The application of the water wheel system has been successfully used at Coventry in a braking device which is attached to the conventional reaction-jet distributors. To obtain extreme low frequency dosing (1 rev in 20-30 min) the most satisfactory method is probably to provide some form of motor-drive.

This could either be attached to the distributor arm and travel round the track as at Ewell, or be stationary and drive through geared wheels on the central distributor column. Such systems could readily include provision for variable speed drives.

To ensure the rotation of the distributors and evenness of distribution under conditions of low flows experienced at some works, the distributors are fed through syphon-operated dosing chambers. It has been suggested that these provide some degree of low frequency dosing, but in practice this does not always appear to be effective, probably because for most of the time the period of application is too long in relation to the intervening rest periods.

Most investigations on the effects of frequency of dosing have been carried out on circular beds and difficulties have been experienced in applying the results to rectangular beds with reciprocating distributors.

The frequency of dosing is affected both by the speed of travel and the distance of travel. Unlike circular beds where the extent of travel - 360°— is fixed, it is possible to design rectangular beds so that at speeds of travel which can be satisfactorily maintained by the driving power provided, the requisite frequency of dosing is achieved.

In fact many of the water-wheel propelled distributors on rectangular beds complete one journey in about 20 min, thus achieving low frequency dosing. With such distribution, however, no provision is available for varying the frequency of dosing and unlike the rotating distributors with staggered jets, they do not provide a higher frequency of dosing within the bed than at the surface.

With motor-driven distributors, by using a longer distance of travel, although economy is effected by the use of fewer machines required, each machine must be capable of distributing a greater flow and the higher speed of travel necessary, results in greater wheel wear per machine.

APPLICATION OF LOW FREQUENCY DOSING TO DOUBLE FILTRATION

According to Howland the contact time in beds is directly proportional to the depth and inversely proportional to the surface hydraulic loading:

$$t \alpha \frac{D}{Q0.66}$$

Thus theoretically, in two-stage filtration of whatever form, where, compared with a similarly loaded single bed, both D and Q can be considered as doubled, the resultant time of contact (t) is increased. The importance of time of contact, or liquid retention time, probably depends upon the nature of the waste; the removal of colloidal and readily physically adsorbed matter being less affected than that of wastes in true solution. Probably more important than the increased retention time with two-stage filtration, is the provision of two distinct environments where organisms responsible for carbonaceous oxidation and nitrifying organisms can exist separately.

Without becoming involved in the controversy whether nitrification occurs in the presence of organic matter, experience with bacteria beds leads one to believe that the autotrophic nitrifying organisms—at least nitrate producing organisms—cannot successfully compete with the more vigorously growing heterotrophic organisms in the presence of available organic nutrient. Thus nitrification is suppressed in the presence of active carbonaceous oxidation, as in the upper layers of a conventional bed where the growth of the heterotrophic bacteria and fungi may produce considerable development of film.

The nitrifying organisms are thus restricted to the depths of the bed where the reduced nutrient concentration limits the development of the heterotrophic organisms. Thus nitrification in the presence of organic matter could be restricted, not by the organic matter itself, but by the suppression of the necessary bacterial flora in competition for oxygen and living space with heterotrophic organisms.

This view is supported by the fact than in A.D.F. the amount of nitrification is much the same in the primary stage, where the organic concentration of the waste is higher, as in the secondary stage. On this basis it was considered that two-stage filtration without alternation would result in a higher degree of nitrification. Previous investigations comparing double filtration with other methods of filtration showed that although under favourable conditions double filtration gave good results, the tendency of the primary bed to clog during the winter made it a less satisfactory process that recirculation or A.D.F.

The success of low frequency dosing in controlling film accumulation suggested that film accumulation in the primary bed of a double filtration plant could be prevented by low frequency dosing. This possibility is being investigated and results so far published of a period including two winters, show that excessive accumulations of film in the primary bed have been prevented.

The 4 hr O.A. and B.O.D. figures for the final effluent were similar to those from similar beds operating on A.D.F at the same overall rate—150 g.y.d. As expected the nitrification was somewhat greater with double filtration, the difference being more marked at higher hydraulic loadings.

Comparisons of different methods of filtration by Peach at Cheltenham and Tidswell at Burton have also shown the superiority of double filtration in nitrification. If low frequency dosing of the primary bed results in double filtration being as efficient as A.D.F. It could be that it is preferable since it would then be possible to differentiate in design and operation between the two beds.

For example the secondary bed could have smaller medium and be dosed at a higher frequency than the primary bed. It would also be possible to have primary and secondary beds of different sizes to enable primary and secondary stages to be carried out at different dosage rates.

HIGH RATE FILTRATION

This term is usually applied to several methods of filtration in which the liquor is applied to the bed at high average hydraulic loadings-500 to several thousand g.y.d. In some cases, as in America, this involves the recirculation of the feed to provide a high dilution ratio. As such it is essentially a growth controlling process. The application of a strong undiluted waste at high hydraulic loadings creates different ecological conditions.

The result of such a practice will probably depend on the nature of the waste and the type of growth it produces under the operating conditions prevailing. With wastes having a high colloidal or suspended organic content, the rate of flocculation probably exceeds that of oxidation and as a result much of the film is composed of flocculated solids. At the high hydraulic rates, the downward rate of flow could remove such unoxidized solids, the bed then acting as a bioflocculation plant, although according to Heukelekian it is still capable of biological oxidation.

If the waste supports a bacterial, as opposed to a fungal growth, it is probable that the high hydraulic rates could bring about periodic sloughing of the film as described by Cooke and Hirsch and Reid and Assenzo. Such sloughing would only occur after considerable accumulation of the film had occurred. When the film is dominated by fungus the resultant tougher growth is able to develop to a greater thickness before being sloughed away by the downward flow of liquid. *Fusarium,* which has tenacious holdfasts, may form the base of such growths on which other organisms become secondarily established. It is when those secondary growths create unfavourable conditions below that the original anchoring growths of *Fusarium* die, and the whole of the film then sloughs off.

In these methods of filtration which rely on the physical scouring of the film, the amount of film present is usually greater than that considered desirable for most satisfactory purification. However, such beds are capable of more than 80 per cent removal of B.O.D. and where a high quality nitrified effluent is not essential, they form a useful economical method of reducing the polluting load of a discharge. They could also be used as pretreatment units before activated sludge plants or bacteria beds.

Although it is not necessary to control film at the low level desirable in conventional beds, the amount present should not impede the free passage of waste or the ventilation of the bed. If ponding results, the downward flow of liquid is impeded and the scouring action reduced. Eventually the film decomposes under the anaerobic conditions so created to form a watery sludge. At such times the efficiency of the bed is greatly reduced and thus the occurrence of such conditions should be prevented. The thickness of film which can be supported without causing such conditions depends upon the size of the voids which is in turn a function of the medium size. Larger voids would also facilitate the passage of the sloughed film through the bed. The solids discharged would probably be composed of larger particles than the humus resulting from the grazing activity of the macrofauna in conventional beds. Investigations on high rate filtration showed that whereas beds constructed of smooth gravel 1 ½ in. to 3 in. in size did not pond seriously when operated at 600 g.y.d., beds constructed of smaller medium ponded continuously during the winter at different rates ranging from 200-1000 g.y.d.

ECOLOGICAL OPERATION

Even in methods of filtration involving nutritional control of the film, the growth rate (R_g) will be positive and has to be countered by a removal rate (R_r). Unless this is achieved, ponding will eventually result and the efficiency of the growth-limiting processes will be further reduced, bringing about a rapid deterioration in conditions. Also the higher instantaneous rates of application, usually involved in such processes, result in more serious ponding if the process breaks down.

In most beds operating at conventional rates, the chief film removal factor is provided by the activity of the grazing fauna. Even high rate beds which normally rely on physical control of film may at times depend upon the activity of grazing fauna for their continued efficiency. In the investigations it was found necessary to allow an *Anisopus* and *Psychoda* population to become established before the beds could be successfully operated.

The establishment and maintenance of an active grazing fauna is thus essential for the successful operation of bacteria beds. The operational requirements necessary to ensure suitable ecological conditions for such a population are a useful guide to bacteria bed operation.

MATURATION

With many wastes, including domestic sewage, the micro-organisms responsible for their breakdown readily become established in the plant, being introduced with the waste water. With specific organic wastes the establishment of the necessary flora may take some time. In such cases the process of maturation can be speeded by inoculating the bed either from a stored culture of the organisms concerned or from a plant treating the same waste.

A satisfactory practical method of inoculating, in the latter case, would be to apply fresh unsettled effluent containing the humus from an appropriate bed, or the effluent from an aeration tank containing activated sludge known to be capable of oxidizing the waste. Since the lag phase of growth of micro-organisms is longer at lower temperatures, an active population of organisms is more rapidly established in the warmer months.

A bacteria bed is often considered matured once nitrification becomes established. Ecologically, however, a bed is only mature when a satisfactory balance between film and grazing fauna has been established. Because micro-organisms are more ubiquitous and more widely dispersed in water and air and because of their rapid rate of multiplication, not only is their chance of being introduced into the bed greater than that of grazing fauna, but their initial rate of colonization is more rapid.

Initially then, there is no effective removal rate (R_r) and until such a force becomes established the growth rate (R_g) of the film should obviously be limited to prevent excessive accumulation of film and ponding. At first little purification of the waste is effected and as a result the concentration of waste at different depths of the bed is much the same and thus the growth rate (R_g) of the film is similar throughout the depth of the bed.

As the developing film in the upper part of the bed removes more organic matter, the concentration available to the film in the depths of the bed reduces the growth rate. The extent to which this control of film within the bed is effected depends on the temperature and hydraulic rate of application, the controlling effect being greater at higher temperatures and low rates of application.

It is thus essential that at first the load to the bed should be severely restricted and that it should only be increased in stages to match the increasing purification capacity of the developing film. The design flow should not be applied until an active grazing fauna population has become established. It is possible that before such a condition is reached, satisfactory purification, even with nitrification, may be obtained.

However, the positive growth rate of the film will eventually bring about a reduction in efficiency unless a grazing population becomes established. Taylor, in reporting his experiences in maturing beds at Bradford, noted that although during the first few months there was a gradual improvement in efficiency, this was usually followed by a marked deterioration for one or two weeks before the beds fully matured.

The time needed for the establishment of the necessary grazing fauna is determined by several factors. Organisms may enter the beds by chance, as eggs introduced by the waste water, or by aerial invasion by adult insects. In the case of beds brought into operation in the vicinity of existing beds, there is every chance of a rapid aerial dispersal to the new beds by flies during periods of suitable climatic conditions.

Also beds constructed on the site of land irrigation areas could similarly be inoculated. Although in time most beds would be invaded by macro-fauna, the process can often be speeded up by the intentional introduction of organisms. The introduction of grazing fauna into beds was widely practised some years ago. *Achorutes* were not only despatched all over this country, but were also exported to places as far afield as Stockholm, Chicago, Singapore and Pretoria. They became popularly known as "Bell's Bugs" after the worker who publicized their beneficial activities in bacteria beds.

The organisms introduced should, of course, be a species suited to the environment imposed by the waste and the operational methods practised. For example with low frequency dosed beds, the introduction of *Lumbricillid* worms would be more suitable than introducing *Achorutes*. Different practical methods are available for effecting their introduction.

Where practicable, medium from a suitable bed containing an abundance of different stages of grazing fauna should be carefully transferred into pockets in the surface of the new bed. On a large scale it is more practicable to apply fresh humus sludge from an appropriate bed via the feed. Although the intentional introduction of fauna may be necessary to speed up maturation, it is more important to maintain conditions to provide a suitable environment in which the desirable organisms can become established. By increasing too rapidly the rates of application before an effective grazing fauna is established, excessive film growth may occur throughout the depth of the bed and the resultant ponding would prevent its successful colonization by the introduced fauna, and its eventual maturation would thus be considerably delayed.

The necessity to severely restrict the loading during the initial stages of maturation can be satisfied in different ways. Probably the most satisfactory way is to operate the beds continuously at very low rates of application. With distributors designed for high flows, such as those on beds operating on double filtration and recirculation systems, it may not be possible to satisfactorily distribute the waste at low flows required. In such cases, beds should be operated at as low flows as possible for part of each day followed by a period of rest. In hot weather this rest period should be during the night when the temperatures are lower, and in cold weather during the day.

In maturing beds designed to operate by the nutritional control of film, although initially the film growth can be encouraged by the conventional operation of the beds, the later stages of maturation should be carried out whilst operating on the appropriate system. This would not only limit the rate of film accumulation but ensure the establishment of the species of grazing fauna characteristic of the system.

SMALL SCALE EXPERIMENTAL BEDS

For experimental purposes and to assess the treatability of wastes, small beds usually consisting of cylinders 6 ft in depth and up to 1 ft in diameter, have been widely used. Within a few weeks it is possible to assess the degree to which the waste is removable. So much can also be assessed by the use of rotary tubes. The suitability of bacteria beds as a method of treatment, however, can only be determined after longer periods when the bed has been fully matured and a film-grazing fauna balance has been established.

With such beds it is usually necessary to introduce the grazing fauna. Care should be exercised in assessing the rates at which the waste should be treated on large scale beds. The size of medium in such beds should, ideally, be such that its diameter is not greater than one eighth of that of the bed.

Account should also be taken of the temperatures at which the beds are operated in relation to those which may be expected in the large scale beds. A further difficulty is in obtaining an effluent sample representative of the final humus tank effluent on the large scale plant. Although satisfactory small humus tanks can be designed for use with such beds, their performance may not be repeated on the large scale plant.

The use of paper-filtered samples ignores the polluting load of an effluent attributable to the suspended solids present. Even for comparative investigations, although paper-filtered samples make comparisons easier, the removal of suspended solids eliminates a variable since the settleability of humus solids produced from beds in different conditions may differ.

ROUTINE OPERATION OF BEDS

Once a bed has been matured, the essence of efficient operation is the maintenance of the balance of populations. To maintain this ecological balance in populations, continuity of operation is essential.

The effect of a period of disuse of a bed depends upon its duration, the original condition of the bed and the climatic conditions prevailing. For some time after the application of the waste to the bed is stopped, the interstitial liquid gradually drains out of the bed for a period depending upon the water-holding capacity of the film.

As the surface layers dry out the grazing fauna retreat into the depths of the bed and soon large numbers of grazing organisms will be observed leaving the bed. The last drainings from the bed are accompanied by vast numbers of the motile stages of insects and worms, migrating from the inactive bed.

The bed is thus seriously depleted of active grazing fauna and presumably the young worms and larvae which hatch from the remaining eggs find conditions unsuitable for development in the drying bed. The pupal stages give rise to flies which, if climatic conditions are favourable, leave such beds in vast numbers. Under such conditions spiders sometimes dominate the scene and whole beds are covered, and the distributors festooned, with their webs. If a bed, having suffered this depletion of grazing fauna, is put back to operate at normal rates of application, serious consequences will probably result.

Such beds, unlike newly constructed ones, are already well inoculated with the necessary micro-organisms, and the film, although possibly having suffered as a result of the period of inactivity, rapidly recovers when the application of the waste is resumed. In the absence of grazing fauna the film may rapidly accumulate and result in ponding which then delays the establishment of a grazing fauna. Such a series of events can result in a bed having a reduced efficiency for long periods, in some cases for over a year.

If the operation of a bed has to be interrupted, then the period of inactivity should be restricted to the minimum and if practicable the bed should be operated for periods whenever possible, even though this may not be convenient. By operating beds during the nights only, it has been possible to carry out major modification over a period of a week or more, without affecting their conditions or efficiency. Under conditions of extreme temperatures, a period of inactivity of longer than one day may have adverse effects; when the weather is mild and wet, longer periods are permissible. When prolonged periods of inactivity are unavoidable, it is then essential to carefully re-mature the bed in a similar manner to the initial maturation process.

Under favourable conditions, usually in the spring, and when flow circumstances permit, the periodic resting of a bed, for periods not exceeding 24 hours, can have beneficial results on a bed in which film accumulation has occurred. Besides the period of starvation imposed on the film, the grazing activity of the macro-fauna is encouraged by the absence of the physical flow of liquid which normally limits their activity. By such a practice the spring unloading, when this is necessary, can be speeded up.

FLY CONTROL

Although ecologically fly populations are best controlled by nutritional control, i.e. by limiting the degree of film accumulation, it may on occasions be necessary to prevent serious nuisance by direct action. Where practicable this may be achieved by flooding the bed and provided that this is of short duration, the film and probably the worm population will not be seriously affected. A more common practice is the application of insecticides. Although these are probably more selective, worms and Collembola not being affected at the concentrations necessary to kill the flies, the bed effluent may be rendered toxic to life in the receiving stream.

For this reason only a small portion of the bed area should be treated at any time. It is usually most practicable to apply the insecticide to the bed as an emulsion or water dispersible powder in the sewage feed. Benzene hexachloride (B.H.C.) at the rate of 1.3 lb of the active gamma isomer per acre of bed has proved successful against *Psychoda* and *Anisopus*. The effectiveness of such treatments is limited by the degree to which the insecticide is evenly distributed to the bed.

Flies which emerge from the beds often congregate in sheltered positions in the vicinity of beds, where they can be attacked by spraying insecticide. This method of attack has least effect on the ecology of the bed and should not, if carefully carried out, result in a toxic effluent. Theoretically the successive applications of insecticides to a fly population in such an isolated habitat might be expected to result in resistant strains being evolved. So far, however, no such case has been substantiated.

Even when insecticide treatments have to be resorted to, they should be restricted to a minimum. A knowledge of the habits of the fly enables insecticides to be used only when nuisance is likely to occur. Fluctuations in the aerial density of *Anisopus* above beds, for example, have been shown to be closely related to climatic conditions. There is also evidence that, although *Anisopus* may be abundant above the beds throughout the summer, they rarely enter near-by dwellings as they do in the spring, and applications of insecticide are, therefore, not necessary after the end of June.

By a continuous succession of insecticide applications it is possible to replace a fly population by one of worms or Collembola which does not give rise to nuisance and which, once the balance of populations has been re-adjusted, is equally efficient in controlling the film.

Once the insecticide treatment ceases, however, the original fauna is rapidly re-established and the continuous use of insecticides, needed to maintain the modified fauna, increases the possibility of a toxic effluent.

Such a change in the fauna is best maintained by controlling the physical environment, as achieved by low frequency dosing.

In this process, not only is the fauna population limited by the nutritional control of its food—the film, but the resultant higher instantaneous rate of application is considered to favour the more prehensile worms. The degree to which this is effective will probably be affected by such factors as the type of distributor jet, jet spacing, type of medium and the hydraulic loading. Where low frequency dosing resulted in such a marked change in the fauna, the beds were operating on A.D.F. and thus the instantaneous rate of application would be double that of a bed operating on single filtration.

DESIGN CONSIDERATIONS

Although a plant has to be designed before it is operated, operational requirements should be the basis of design. The operator should not find himself severely restricted in the degree of control by over rigid design based on arbitrary criteria. For this reason, the brief discussion on the ecological factors which should be considered in design of beds has been left to follow the operational considerations.

SITE

Because of the adverse effect of low temperatures, beds should be sited in sheltered positions. Enclosed beds should prove advantageous in this respect. Shelter provided by a screen of trees can, however, cause a nuisance at leaf-fall. Temperatures within the bed are largely determined by the temperature of the waste and in this connection the proximity of the plant to the source of waste may be important, reductions in sewage temperatures may occur in long trunk sewers, for example.

Because of the possibility of fly nuisance beds should be at some distance from dwellings, etc., especially those to the leeward in relation to the prevailing winds. Hydraulic considerations are paramount in the design of most plants and the choice of site is obviously important in this respect. In some situations careful siting can enable double filtration to be practised without additional pumping.

MEDIUM

The function of the medium is to provide the necessary surface on which the active biological film can be supported and at the same time permit the free flow of waste and air to the film. The size, shape and surface nature of the medium are important features determining the degree to which these functions are carried out and hence affect the efficiency of a bed.

Theoretically it appears wasteful in plant capacity that, to provide the necessary surface area, more than half the volume of a bed constructed of conventional medium is occupied by inert matter—the medium. In America several attempts have been made to construct suitable media in units of uniform size, e.g. "Aero-blocks" "Raschig rings", and "Straight's blocks".

Goldthorpe in this country designed the "Huddersfield Tile". The more recent developments of plastics have permitted the prefabrication of corrugated plastic surfaces as bed medium, trade registered as "Dowpac. Although with this medium the surface area per unit volume of bed is no greater than that of stone media, only 6 per cent of the bed volume is occupied by the material. A development of this type of medium to increase the surface area should further increase its efficiency.

The extent to which the grazing fauna could become established on such medium is not known and although bacterial film may be removed hydraulically the possibility of clogging by profuse growths of fungus should not be overlooked. Until a satisfactory and economic alternative is found, the conventional medium will continue to be used in most cases.

Since the type of medium used may affect the ultimate efficiency of the plant, careful consideration should be given to its choice. It has been shown that the *liquid* retention time of liquids flowing over spheres varies inversely with the two-thirds power of the diameter of the spheres. Thus, medium size could be expected to affect liquid retention time.

More important, however, is the effect of medium size in determining the amount of film which a given volume of bed can support. The optimum size of medium is determined by two opposing factors; the smaller the medium, the greater the surface area, but the more restricted the interstitial spaces in which the film can accumulate and through which the waste and air can flow. The optimum size of medium is thus the smallest on which the maximum degree of film accumulation, resulting from the conditions of operation, can be accommodated without interfering with the ventilation of the bed or the even distribution of the waste. Since the degree of film accumulation depends upon several factors, including the nature of the waste and method of operation, the optimum size of medium is affected by such factors. It is thus unlikely that there can be a single standard medium for all beds, considering the wide range of conditions under which they are operated. Since, at least with conventional beds, the degree of film accumulation is greater in the upper layers, on this basis smaller medium could be used in the depths of the bed.

However, the amount of humus in the liquor increases as it passes downward and thus very small medium would be unsuitable. With low frequency dosed beds in which the load is more evenly distributed the size of medium throughout the bed should be the same, providing the lower-most medium is sufficiently large to prevent it passing through the openings in the under-draining. Where the degree of film accumulation can be limited, the use of smaller medium is possible in such beds.

The base of the bed should have sufficient slope to prevent the accumulation of humus solids under the tiles which could impede the ventilation of the bed. The presence of "fines" of appreciably smaller size than that considered suitable, even in relatively low proportions by weight, is undesirable. Medium of mixed sizes tends to become consolidated and the interstitial spaces between the suitable medium become restricted by the smaller components. For the same reason the media used should not be susceptible to disintegration and should be handled carefully when filling the beds.

As important as size, is the shape of individual units of medium. Ideally, they should be almost spherical. Excessively long or flat particles tend to pack and restrict the interstitial spaces.

The nature of the surface of the particles is another factor to be considered. Bacterial slimes and fungal growths can develop on smooth surfaces, such as washed gravel, although the initial lag phase may be somewhat longer. It is probable, however, that a rough surface such as cracked granite, provides a better holdfast for certain grazing fauna. The honeycomb-like structure of slag or clinker is ecologically superior to other standard types of medium. It presents a larger surface area than other media of the same size, even though all of the surface may not be wetted by the waste.

Furthermore, it provides an ideal niche for the grazing fauna. One has only to examine a matured piece of slag medium and see the worms, larvae and blue Springtails housed in their ready-made caves and burrows and grazing on the film growing on their doorsteps, to appreciate the hospitable environment it provides. Economic considerations will undoubtedly be a major factor in the choice of medium.

It may be more economical, for example, to use suitable locally available material, even though this may necessitate having slightly larger beds than would be required if a more costly superior form of medium were used. This consideration, however, only applies to the nature of the medium and not to its grading. To use cheap media of indifferent grading containing "fines", or media that will disintegrate to produce "fines" is certainly false economy. Such media will most probably give rise to trouble under extreme operating conditions which could only be remedied at considerable expense and inconvenience.

DEPTH

Little consideration has been given to the question of bed depth, and by tradition, most beds in England are approximately 6 ft deep. The depth of the bed, however, is ecologically important in as much as, at a constant loading (G.Y.D), it affects the surface rate of application, and thereby the downward rate of flow of the waste.

In conventional beds with high frequency dosing, the increase in surface loading with increase in bed depth could cause excessive accumulations of solids near the surface of deep beds which would reduce their efficiency. With low frequency dosing, however, the more even distribution of the film throughout the depth of the bed should make possible the use of beds considerably deeper than 6 ft.

Where this is possible it would reduce the land area required and the size of base and under draining. The numbers of distributors would be reduced but their capacity would need to be greater.

The retaining walls, however, become increasingly more expensive with increase in height above ground level; this being more so with rectangular beds than with circular ones. Where hydraulic requirements demand, it will be necessary to provide pumping capacity together with an adequate standby pump.

In two stage filtration or recirculation, where pumping is only required for the primary or returned effluent, it is often possible to continue to apply the waste by gravity to the beds when pump failure occurs. This would not be possible in the case of a deep (high) bed where pumping was normally required. Theoretically on the basis of the formula:

$$t \, \alpha \, \frac{D}{Q^{0.66}}$$

where D = Depth and Q = Surface loading, by increasing the depth and proportionately increasing the surface loading to maintain the same G.Y.D. loading, the liquid retention time is increased.

SHAPE

Although the old method of fixed spray distribution allowed full use to be made of any shape of available area, modern methods of distribution usually determine the bed shape as rectangular or circular. Other shapes demand complication in the design of distributors. Rectangular beds are usually only found on the larger works although some large works have circular ones.

Ecological differences occur because of the different distribution methods imposed by the different shapes. The rectangular beds, although more readily ensuring an even distribution of waste across the width of the bed, are not so readily adapted for low frequency dosing.

VENTILATION

Although ventilation is essential for the successful operation of beds, the method by which it is effected is not fully understood. It is probable that several forces are at play in effecting ventilation, and the extent to which each contributes varies under varying operating conditions.

In most beds, at most times, the temperature differences between the interior of the bed and the air above the bed results in convection currents which effect ventilation. It has been calculated that a difference of 1°F is sufficient to give adequate ventilation to a bed treating sewage at normal rates. The importance of maintaining a thin film so as not to impede the ventilation currents and to ensure that all the organisms in the film receive an adequate supply of oxygen has already been discussed. Structurally, provision should be made for the ventilation of the bed from below. With circular beds having under drains discharging into a circumferential open channel, these should provide adequate ventilation, providing that they are not surcharged.

The use of open stone walling for such beds is probably not essential and it certainly provides an attractive shelter for flies. Where the under drains discharge into one small common effluent channel within the bed, the provision of ventilation shafts on the periphery of the bed would seem to be a wise measure. With rectangular beds having the under drains discharging independently to a longitudinal channel to the outside of the bed, or to a channel in a well ventilated duct within the bed, the under drains should themselves provide adequate ventilation providing that they are kept open.

From an aeration point of view beds cannot be over ventilated, although in the winter this could lead to an undesirable reduction in bed temperatures. Since in most beds ventilation is probably not a limiting factor, the provision of forced ventilation has not always proved beneficial. It could possibly increase the efficiency of enclosed beds and those treating strong wastes.

Distribution of Waste As we have seen, the method of distributing the waste can markedly affect the ecology of the bed and thereby its efficiency. Distributor design is an integral part of bed design and should be considered in relation to the proposed method of operation. Difficulties arise in providing for even distribution of the waste over a wide daily range of flows experienced at some works. At very low flows recirculation of effluent is recommended.

On the whole, if even distribution over the whole flow range is unattainable, the distributor should be designed to ensure even distribution at average flows and above, rather than at very low flows. This being so, the uneven distribution at low flows can be regarded as producing a greater degree of under loading in some places, but the uneven distribution at high flows can only result in a greater overloading in some places—a much more serious effect.

The view that a dribbling jet is likely to produce more growth than a normal jet is based on the erroneous mechanistic concept of film control. Apart from the possibility of freezing in winter, and provided that the distributor is in motion, dribbling during periods of low flow should have no adverse effect. The provision of motor drives for distributors ensures movement at times of low flow besides providing for controlled frequency dosing and makes unnecessary the provision of syphon dosing chambers. With normal jets the spacing between them should be such that at average flows the distance between jets lines made by all the distributor arms serving one area of the bed are not greater than 6 in.

On circular beds, to ensure even distribution along the radius, some distributors have jets of the same size spaced at increasing distances towards the centre of the bed. Even distribution, however, is probably better achieved by having the jets equally spaced and having a progressive reduction in the diameter of the jets towards the centre of the bed. In beds where nutritional methods of film control are practised, the use of more widely spaced jets with splash plates, to ensure even distributions, should be equally satisfactory.

By spraying or cascading the waste from the distributors, the increased aeration should be beneficial but the amount of oxygen so dissolved can only satisfy a fraction of the total oxygen demand. Fixed spray-jets, because they provide a continuous rain of waste on the surface of beds, encourage excessive growths of film and most have now been replaced by travelling distributors giving a lower frequency of dosing.

They could, however, be used on the secondary bed of a double filtration unit, where the low nutrient concentration should not result in excessive growths of film. Such a method of application would probably enhance nitrification within the bed. Because of the need to ensure continuity of operation, distributors independently driven are superior to those rope-hauled in groups. Even when use is made of the provision for uncoupling a faulty distributor for attention, when this is possible, the area served by it receives no sewage, whereas it is sometimes possible to permit an independently driven distributor to cover the extra area by extending the length of travel.

ANCILLARY PLANT

Beds should not be used to remove settleable solids and it is usually necessary to provide adequate settlement of the waste before it is applied to the beds. It is equally important to remove the humus solids discharged from the bed before the effluent passes to the receiving stream. The conventional method is to settle these solids in humus tanks; deep tanks of small cross sectional area which provide a fairly even rate of upward flow are probably the best type. The humus sludge should not be retained for any length of time. If nitrification is occurring in the bed, the denitrification, which would rapidly occur in the anaerobic conditions in the tank, could liberate nitrogen gas which, in rising, would carry the sludge to the surface.

The sludge should be removed at least once or twice each day if not continuously. The monthly washing down of humus tanks, where practicable, should also prove beneficial. Humus from beds differs considerably, depending on the waste being treated and the condition of the bed, some having better settling qualities. However efficient the humus tanks, some finely divided solids are, in practice, always present in the effluent and these form a considerable portion of the oxygen demand of many effluents. Their removal can be effected by sand filtration or micro strainers. Where land is available the irrigation of the effluent over a grass plot is a simple and effective method of removing most of the solids and effecting clarification when necessary. Such grass plots should be used for the removal of the finely divided solids which do not settle in humus tanks, and not to replace humus tanks or supplement grossly inefficient tanks, unless the frequent removal of the accumulated solids from the plots is anticipated.

Chapter 11

Disruption of Ecological Systems

The direct effects of pollution on property, on human health, and on the quality of life are varied and important, but they may ultimately prove to be less critical for society as a whole than the less obvious effects of pollution and other human activities on the ecological systems that sustain human life.

Crucial to these systems are the green plants that are the basic energy source for all other forms of life on Earth, and green plants are provided with the material ingredients of growth by the nutrient cycles of the biosphere. Those cycles do routinely what human society as yet cannot do — convert wastes completely into resources, using energy from the sun.

The crops of civilization are watered, nourished, and protected from potential pests with considerable help from natural processes, which include the formation and preservation of soil itself. The agents and carriers of human disease, like crop pests, are controlled more often by natural enemies or by environmental conditions than by human action. Climate and the composition of the atmosphere itself are regulated by geophysical and biological processes that may be susceptible to human interference. These and other "public services" of the global ecosystem cannot be replaced by technology now or in the foreseeable future. This is so in many instances because a process is not understood scientifically. And, even where scientific knowledge might be adequate, the sheer magnitude of the tasks performed by natural ecosystems dwarfs civilization's capacity to finance, produce, and deploy the technology necessary to replace them.

Since the support provided for human life by ecosystems is indispensable and irreplaceable, it is essential that the mechanisms behind those vital functions — and the potential for disrupting them through pollution and other human activities be carefully studied.

MODIFYING ECOSYSTEMS

Like any other creatures, human beings have always had some

effect on the ecological systems of which they were a part, but the history of deliberate and significant modification of ecosystems doubtless began with the development of agriculture. Since then, human influence has reached to the remotest areas of the planet and has been intensified by increasing numbers of people and the development of modern technology.

What are some of the ways in which humanity modifies ecological systems? Obviously, some ecosystems are destroyed outright by such diverse activities as clearing land to plant crops, starting fires, building dams, applying defoliants to jungles, constructing buildings, and laying pavement. Forests are chopped down for structural wood and firewood. Estuaries are dredged, filled, or altered by bulkheads.

The effects of such assaults are diverse. Paved areas are normally removed from natural systems for long periods — and their rapid reclamation is both costly and difficult. Dams take land out of terrestrial systems for long periods, but as noted earlier, dams are temporary structures, and enriched land will eventually be returned.

Strip-mined areas in developed countries often have not been reclaimed; the practicability of doing so at all is a function of both economics and ecology. When prairie is converted to cornfield, an unstable ecosystem replaces a stable one. Attempts to stabilize such systems artificially can lead to further destabilization and bring about changes elsewhere as well.

Physical assaults on ecosystems often interact with over exploitation and pollution to threaten species with extinction. For example, industrial and harbor development encroaching on breeding areas is combining with oil pollution and competition from the fishing industry to deplete South African jackass penguin populations already reduced in size by over exploitation of the bird's eggs and disruption (since halted) of their breeding activities. Now that ecologists are gaining a more detailed understanding of the factors that threaten ecosystems, they are beginning to examine carefully strategies for preserving natural diversity. Especially interesting have been attempts to develop a guideline for location, number, size, and configuration of natural reserves, based on the theory of island biogeography.

Few topics are more worthy of investigation, although, of course, application of the principles uncovered will require a commitment to conservation and a halt to human expansion while there are significant wild areas left to conserve.

Deforestation

The values to human society of healthy forest ecosystems extend far beyond the obvious ones of timber, wood products, and recreational resources. Forest ecologist F. H. Bormann has summarized these additional less appreciated, but perhaps ultimately more important values: "Forest ecosystems... exercise considerable control over patterns of climate, hydrology, circulation of nutrients, erosion, and the cleansing functions of air and water, as well as over the status of streams, lakes, and underground water supplies."

Moreover, tropical forests especially represent a vast reservoir of genetic diversity of plants, animals, and microorganisms, any of which potentially could be of direct benefit to humanity, and all of which play their parts in maintaining the ecosystem.

Wholesale logging of forests creates extreme changes that effectively destroy the forest ecosystem. Numerous animals that depend on the trees for food and shelter disappear. Many of the smaller forest plants depend on the trees for shade; they and the animals they support also disappear.

Removal of temperate-zone forests also results in heavy losses of nutrients from the soil. These losses may overload streams with nutrients and, if prolonged, may limit the ability of succession to restore the forest ecosystem. If disruption of a forest floor is severe, widespread erosion losses may become serious. If such disruption is avoided, the potential for recovery remains high if the successional process can begin within a year or two of cutting, since early successional species are adapted to take advantage of the increased abundance of water and nutrients in such areas.

In time, if reforestation is not carried out, a deforested area is invaded by pioneer weedy plants that often have far fewer desirable characteristics for human use than did the forest that was removed. Areas from which tropical forests have been removed may have much less potential for successional recovery, particularly where soils are lateritic.

Loss of plant cover and topsoil reduces the capacity of an area to retain water, diminishes the supply of fresh water, causes silting behind dams, and has other serious consequences for human beings. The flooding along many of the world's rivers, from the Ganges in India to the Eel River in California, has been greatly aggravated by heavy deforestation in their watersheds.

Deforestation also significantly reduces the amount of water transferred from ground to air by transpiration by trees, which in turn affects climate, at least locally. Downwind of a deforested area, weather is made more arid and subject to greater extremes of temperature.

Removal of tropical rain forests also reduces the absorption of incident sunlight. As tropical forests are progressively cleared during the rest of this century, the result may be unanticipated changes in climate, perhaps on a global scale. Indeed, some changes attributable to tropical deforestation have been noted already.

Human activities have already substantially reduced the world's forests and produced a great increase in the amount of desert and wasteland. Such destructive activities have accelerated as the human population has increased, and in many cases complex interactions between human beings and the environment have led, or have threatened to lead, to *ecocatastrophes* (dramatic failure of the life-support capability of an ecosystem).

Slash-and-Burn Agriculture

One of the clearest examples of how overpopulation can lead to environmental deterioration can be seen in one kind of non-mechanized farming. In tropical forests, people for centuries have practiced what is known variously as slash-and-burn or milpa or swidden agriculture. In that method a clearing is cut in a forest during the driest part of the year and the felled trees left in place. After they have dried, they are burned.

The fire not only provides fertile ash, it also kills insects and the seeds of weeds. The farmer plants a crop as rain commences and then injects additional energy into the system in the form of labour fighting weeds with hoe and machete.

After the harvest the inedible parts of the crop plants are chopped down, allowed to dry in the field, and burned. Since the mass of plant material is smaller than that of the previous year when the forest was first cleared, the fire is not so hot and there is less ash. As a result the soil is less thoroughly fertilized and the pest kill is less complete.

The downhill spiral continues for a year or two until it is no longer profitable to farm that clearing. The milpa farmer then moves on, creates a new clearing, and repeats the whole process. Meanwhile the forest gradually reclaims the original clearing. Tens or hundreds of years later, the original clearing may once again be farmed.

Because the tropical forest is a tightly closed system as far as nutrient cycles are concerned, the small, widely spaced clearings of swidden farmers do little damage to that system. Adjacent trees help protect a clearing from the torrential tropical rains, and their roots, extending into the clearing, recapture nutrients. When a clearing is abandoned, the forest can reinvade it.

But what happens when human population growth forces farmers to return to previously farmed plots before the forest has totally reclaimed them? Or when clearings are made larger in order to support more people?

When a sizable chunk of forest is cleared for agriculture (or defoliated for military purposes), the continuous recycling of nutrients is interrupted. The surrounding forest no longer breaks the force of the rain, and tree roots penetrate only the periphery of the clearing.

Heavy rains wash away the thin supply of soil nutrients, of which the last to go (where they are present) are iron and aluminum oxides. When certain iron-rich lateritic soils, which underly perhaps 5 to 10 per cent of tropical forests, are eroded and exposed to sun and oxygen, a series of complex chemical changes may take place, resulting in the formation of a rocklike substance called laterite.

Laterization has occurred over wide areas of the tropics, starting long ago and continuing in recent years. Those who have been fortunate enough to visit Angkor Wat in Cambodia have seen magnificent cities and temples built by the Khmers 800 to 1000 years ago. The construction materials were sandstone and laterite, and laterization may have been a principal reason for the disappearance of the Khmer civilization shortly after its greatest monuments were constructed. After laterization, unfortunately, the forest normally is unable to reinvade a clearing successfully. Even without laterization, a cleared area will rapidly lose the nutrients required for satisfactory agricultural yields — nutrients that can only be restored by forest regrowth or by careful fertilization.

Fig. Laterite Construction in the Temple of Angkor

Catastrophic loss of soil fertility is thus a natural consequence of increased population density among slash-and-burn agriculturalists. In order to support more people, the age-old patterns of movement are accelerated.

Sites of previous clearings are revisited before the forest has totally reclaimed them. As a result, the fires are less hot, the ash less abundant, and soil fertility decreased. Clearings must be abandoned ever more often, and, to compensate, larger clearings are made. The road to soil deterioration is open.

The general pattern described has been repeated in many tropical areas when and where the human population has increased. It has been hypothesized that over intensive tropical agriculture contributed to the collapse of both the Khmer and the classic Mayan civilizations — although it seems likely that both of those societies had advanced to permanent, rather than milpa cultivation before laterization and other problems overtook them.

Farming the Tropics

A traveller in the tropics today cannot help being impressed by the poor quality of much of the land cleared of forest there. At best, a great deal of that land can serve only as second-rate pasture.

Silting of rivers is also easily observed, but subtle changes in climate resulting from removal of forest are less obvious.

Tropical forests are usually able to reinvade small clearings where soils are subject to laterization before the process is complete. Whether large areas that have been kept cleared for substantial periods of time can be reforested after laterization is an open question.

Reforestation has been successful in some areas where the soil crust has been carefully broken up, fertilizers applied, and the entire system carefully cultivated. But natural reforestation (and, especially, reestablishment of a semblance of the original diversity) seems unlikely. Even most of the attempts supported by human activity have failed. Laterization is continuing in various parts of the tropics and will doubtless proceed more rapidly as people become increasingly desperate for food.

"The ambitious plans to increase food production in the tropics to meet the pressure of the rapid rise of population have given too little consideration to the laterization problem and the measures that will have to be undertaken to overcome it." She points to one fiasco at Iata in the Amazon basin, where the government of Brazil once attempted to found a farming community. Laterization destroyed the project, when "in less than five years the cleared fields became virtually pavements of rock."

The degree of threat to tropical agriculture posed by lateritic soils is a matter of controversy. Unfortunately, tropical soils are not well understood, so prediction of the consequences of clearing tropical forests for agricultural purposes is difficult.

Some things can be said with assurance, however. First of all, the term *the tropics* covers a multitude of sins. Far from being uniform, the tropics are more diverse ecologically than the temperate regions. Therefore, the ecological and agricultural situation for each location must be evaluated separately. And, furthermore, that evaluation may not provide quick and easy solutions to the problems of agricultural development.

For instance, much of the Amazon Basin presents a severe challenge to agriculture. With great effort and care, successful agricultural programs could be developed to farm the so-called *terra firme* — land above the floodplains that are periodically inundated (and thus fertilized) by the Amazon. Such programs would involve the interlocking of farming with animal husbandry, to make manure available for fertilizer (it tends to be much more satisfactory under tropical conditions than inorganic fertilizers).

The constraints on agriculture in Amazonia, moreover, are such as to make difficult the continual adjustments necessary to meet changing local and international market conditions.

THE TRANS-AMAZONIAN HIGHWAY

In 1970, Brazil began a crash programme of highway construction in the Amazon basin. The central project is the Trans-Amazonian Highway. When completed, the highway is to connect such Atlantic coast ports as Belem and Recife with points in Peru and Bolivia, through which countries they may someday be linked by road to the Pacific.

The route of the Trans-Amazonian Highway runs roughly parallel to the river itself, about 320 kilometers to the south. The highway is planned to tie in with an extensive additional network of roads crisscrossing the basin.

The programme was initiated by President Emilio Médici after he had visited the drought-devastated northeast of Brazil in 1969, in order "to carry the landless population of the Northeast to the unpopulated lands of the Amazon." An agritown of forty-eight to sixty families is planned for every 10 kilometers along the 4800 kilometer length of the highway. The government has expropriated all land within 50 miles of the road network (an area of almost 2.3 million square kilometers — much larger than Mexico), on which it plans to establish colonists on 100-hectare (250-acre) parcels.

Some 8000 Amazonian Indians are to be resettled from the development area onto reservations in a move reminiscent of United States policy in the nineteenth century. Although eighty-four Brazilian social scientists signed a protest against the plan for the Indians, it is clear that the development project will be permitted to move ahead.

The great Amazonian rain forest, complete with its rich array of animals and peoples, appears destined to follow the tall-grass prairies of North America to extinction. But the tragedy of the Amazon will be even greater. The European successors of the North American Indians at least were able to extract great wealth on a continuing basis from the lands they appropriated. But the current efforts of the Brazilians will, in all probability, merely net them a few harvests at the cost of irreversible destruction of the region. Amazonia could be turned into the world's largest parking lot.

The tragedy is all the greater because there have been ample warnings for Brazilians — warnings from home, not just from foreign ecologists. The failure of the agricultural colony set up at Iata was one portent. But the Brazilian government, like other governments, seems to find it hard to learn from experience.

And, like other governments, it also may turn a deaf ear to expertise when politics and special interests are involved. Scientific reports published by a government agency, Instituto Brasileiro de Geografia, pointed out the pitfalls of the development programme, warning that "a disaster of enormous proportions" would occur if the policy were carried through.

The key to the situation could be the rumor that groups behind the development were actually interested in opening the Amazon basin for oil and other mineral exploration — that agriculture was a secondary concern, more of an excuse than anything else. But it is not certain that the forest covers a mineral bonanza.

It is possible that Brazilians do not really want to know the truth about Amazonia, for that would force them to confront the true dimensions of their country. Gone would be myths of untold abundance, of limitless resources, of future world power.

Brazilians would have to recognize the biological and physical limits within which their nation must operate in order to survive -much as citizens of the United States in the 1970s have had to face the limitations to their power. But the Brazilians have a tougher job, since they must confront reality *before* they have had their fling at being overdeveloped. And there is little sign that they are ready for that corifrontation.

In Brazil today, one sees a striking example of a ubiquitous phenomenon — the inability of social systems to respond effectively to imperatives arising from the laws of nature. Since the laws of nature are unlikely to be repealed, it would behoove humanity to see whether the "laws" of social behaviour are equally fixed. If they prove to be, as some social scientists seem to think, then the outlook for *Homo sapiens* is bleak indeed.

DESERTIFICATION AND THE SAHEL FAMINE

Another dramatic example of the interaction of human population size with natural systems is seen in the Sahel famine. The climatic events triggering the famine but people themselves played a major role in creating conditions in which famine could be triggered. In some areas, for instance, colonial officials misinterpreted the ecologically sound pastoral systems of nomads living in the arid region as inefficient and wasteful.

They limited the nomads' movements and attempted to commercialize the economy so that meat could be sold. Following the drilling of thousands of wells, the building of storage tanks for water, and the introduction of veterinary medicine, the herds of cattle in the area were greatly increased.

The inevitable result was overgrazing, soil erosion, and desertification — since the removal of vegetation greatly increases the rate of runoff and thus reduces the beneficial results of rain. As ecologist J. L. Cloudsley Thompson wrote, "Man cannot escape responsibility for having created a large part of the Sahara desert.

In regions where the needs of the vegetation are delicately balanced by precarious rainfall, overgrazing and the utilization of firewood can have a disastrous and irreversible effect."

The populations of human beings and domestic animals in the Sahel were outstripping the area's ability to produce food and forage before the drought of the late 1960s and 1970s. The repeated failures of the monsoon in the early 1970s were the final blow. The herds of many pastoralists were virtually wiped out.

Hundreds of thousands — perhaps as many as 1.5 million people had died of hunger by the end of 1974. Because the population is scattered and government figures untrustworthy, the true dimensions of the disaster will probably never be known. In Ethiopia, for instance, deliberate attempts were made to cover up the famines.

The way of life of many of the pastoral peoples seems to have been destroyed. Nomadic peoples like the Tuareg, the fabled "blue men" of the desert, have crowded into hideous slums on the margins of poverty stricken towns in the southern Sahel. There, on relief, they are turning to farming to supplement their meager rations. Suggested cures for the problem indicate that no lessons have been learned. The president of Niger has urged, for instance, that international aid be supplied to drill 2500 wells, averaging 300 meters deep, across that country. The cost would be almost a quarter of a billion dollars, and successful completion of the project would almost certainly result in an acceleration of desertification through overgrazing.

Agriculture and Instabilities

Even where overpopulation, the quality of soils, lack of rain, erosion, or salt accumulation do not pose special problems, the practice of agriculture alone may cause ecological difficulties. The most basic one is that agriculture is a simplifier of ecosystems, replacing relatively complex natural biological communities with relatively simple man-made ones based on a few strains of crops.

Agricultural communities thus tend to be less stable than their natural counterparts: they are vulnerable to invasions by weeds, insect pests, and plant diseases, and they are particularly sensitive to extremes of weather and variations in climate.

Historically, humanity has attempted to defend agricultural communities against the instabilities to which they are susceptible by means of vigilance and the application of "energy subsidies" (for example, by hoeing weeds, and more recently by using pesticides and mechanical cultivation). The effort has not always been successful.

Any monoculture is highly vulnerable to plant pests and diseases. An insect pest, a fungus, or a strain of bacteria can sweep through the entire plant population. A classic example of the destruction of a monoculture is provided by one of the first Green Revolutions — the introduction of the potato into Ireland in the eighteenth century.

The potato provided the basis for an increase in the Irish population from 3 million people living in poverty to 8 million living in poverty. In such situations, expansion of the food supply does not necessarily lead to an increase in the standard of living; more frequently, it leads only to more people living at the same low standard. Between 1845 and 1848, a fungal blight invaded the fields, producing one of history's great disasters — the Irish potato famine.

The unstable potato ecosystem collapsed and with it the artificially augmented carrying capacity of the land. In four years of crop failures, 1.5 million people starved to death, 2 million emigrated, and millions of others underwent great suffering through semistarvation. Advances in agricultural technology in the past hundred years have not resolved the ecological dilemma of agriculture; they have aggravated it.

The dilemma can be summarized in this way: agriculture tries to manage ecosystems in such a way as to maximize productivity; nature "manages" ecosystems in such a way as to maximize stability. The two goals are incompatible. In short, in agriculture productivity is achieved at the expense of stability.

Of course, humanity must practice agriculture to support even a fraction of the existing human population. A degree of instability in agricultural ecosystems must therefore be accepted and, where possible, compensated for by technology. However, the trends in modern agriculture, associated in part with the urgent need to cope with unprecedented population growth and in part with the desire to maximize yields (the amount of harvest per hectare) for strictly economic reasons, are especially worrisome ecologically. The three major liabilities are:

- As larger and larger land areas are given over to farming, the unexploited tracts available to serve as reservoirs of species diversity and to carry out the public service functions of natural ecosystems become smaller and fewer.
- Even in parts of the world where land area under agriculture is constant or (for economic reasons) dwindling, attempts to maximize yields have led to dramatic increases in the use of pesticides and inorganic fertilizers — increases that have far-reaching ecological consequences themselves.
- The quest for high yields has led also to the replacement of myriad traditional crop varieties around the world with a few, specially bred, high-yielding strains. Areas of unprecedented size are now planted to single varieties of wheat or rice. Because of this enormous expansion of monoculture, the chances are much greater for an epidemic crop failure from insects or disease — one that might occur over an area far greater than any crop failure in the past.

POLLUTANTS IN ECOSYSTEMS

The expansion and intensification of agriculture has been accompanied by a continuing Industrial Revolution that has multiplied many times over both the magnitude and the variety of the substances introduced into the biological environment by human beings. Substances as either *qualitative pollutants* (synthetic substances produced and released only by human action) or *quantitative pollutants* (substances naturally present in the environment but released in significant additional amounts by human activities).

Well-known qualitative pollutants include the chlorinated

hydrocarbon pesticides such as DDT, the related class of industrial chemicals called PCBS, and some herbicides. These substances are biologically active in the sense that they stimulate physiological changes. But, because organisms have had no experience with them over evolutionary time, the substances are usually not biodegradable. Indeed, some, like the pesticides kepone and mirex, are among the most stable molecules ever synthesized by chemists.

Thus, they may persist in the environment for years, and even decades, after being introduced, and be transported around the globe by wind and water. Their long-term effects will be discovered only by experience. The significance of quantitative pollutants in a given situation must be judged according to whether the natural abundance of the agent is altered slightly or substantially, locally or globally, suddenly or gradually, and according to how harmful the agent is at its naturally occurring concentrations.

Concentration of Toxic Substances in Ecosystems

Nowhere is human ecological naiveté more evident than in popular assumptions about the capacity of the atmosphere, soils, rivers, and oceans to absorb pollution. Those assumptions all too often take the following form: if 1 liter of poison is added to I billion liters of water, then the highest concentration of poison to which anything will be exposed is about one part per billion.

This might be approximately true if complete mixing by diffusion took place rapidly (which it often does not) and *if only physical systems were involved*. But because biological systems are involved, the situation is radically different. Ignorance of this point is not confined to laymen. For instance, Sir Robert Robinson, Nobel laureate in organic chemistry, made the following classic blunder.

In a letter to the London *Times* discounting the threat of leaded compounds to oceanic plankton, he began: "Neither our 'Prophets of Doom', nor the legislators who are so easily frightened by them, are particularly fond of arithmetic..." and then went on to do some "simple arithmetic" to show that the dilution of lead in the oceans would be so great that the lead would remain biologically negligible. Of course, his whole exercise completely missed the point, because of — among other things — the phenomenon of biological amplification.

Filter-feeding animals, for example, may concentrate poisons to levels far higher than those found in the surrounding medium. Oysters capture food by constantly filtering the water they inhabit, and they live in shallow water near the shore, where pollution is heaviest. Consequently, their bodies often contain much higher concentrations of radioactive substances or dangerous chemicals than the water in which they live.

For instance, oysters have been found to accumulate up to 70,000 times the concentration of chlorinated hydrocarbon insectides that exists in their environment. Food chains may also lead to concentration of toxic substances; they act as a kind of biological amplifier. The diagram of a Long Island estuary food web how the concentrations of DDT and its derivatives tend to increase in food chains from one trophic level to another.

That tendency is especially marked for the chlorinated hydrocarbons because of their high solubility in fatty substances and their low solubility in water. Although the clam and the mud snail are at the same trophic level, the filter-feeding clam accumulates more than half again as much DDT as the mud snail because of the difference in their food-capturing habits.

The mechanism of concentration in food chains is simple. One would expect from the second law of thermodynamics that the mass of herbivores normally is not as great as the mass of plants they feed on. With each step upward in a food chain the biomass is reduced.

Energy present in the chemical bonds of organisms at one level does not all end up as bond energy at the next level, because much of the energy is degraded to heat at each step. In contrast, losses of DDT and related compounds along a food chain are small compared to the amounts that are transferred upward through the chain because they tend to be neither degraded nor eliminated from the organisms. As a result, the concentration of DDT increases at each level. Concentrations in the birds at the end of the food chain are from tens to many hundreds of times as great as in the animals farther down the chain.

Food chain concentration in aquatic environments is not as frequently observed as in terrestrial ones because of the tendency at each trophic level for the concentrations of compounds to equilibrate between organism and water — the equilibrium being roughly the same at all levels. Concentration will only occur if movement from level to level is fast enough to prevent the reaching of equilibrium at each level.

Insecticides in Ecosystems

Besides being direct threats to human health, synthetic insecticides are among society's most potent tools for simplifying and thus destabilizing ecosystems. The increase in the concentration of the most persistent of those compounds with each upward step in a food chain exposes the populations least able to survive poisoning to the highest concentrations. There are several reasons why the organisms near the upper ends of food chains are less able to cope with the poisons than are, for example, herbivores. The first traces back again to the second law of thermodynamics. Because of the loss of available energy at each transfer, the higher a given population occurs in a food chain, the smaller that population is (everything else being equal).

This means that if a poison were applied that would kill most of the predators and herbivores in an area, it would be more likely to exterminate the population of predators than the population of herbivores, simply because there would be fewer predators to begin with. Purely by chance, some members of the larger population would be more likely to survive.

It is not even necessary to kill all individuals of any one species of predator to force it to extinction. If survivors were too scattered or weakened for the sexes to find one another and produce offspring, extinction would surely follow. Or, if survivors were few, loss of genetic variability might cause the population to dwindle to zero.

Another reason that small populations of animals high on a food chain are more vulnerable is that smaller populations tend to have less genetic variability. Suppose, for instance, that I individual per 100,000 in each of two species of insects by chance carries a mutant gene that makes it naturally resistant to a certain pesticide.

Assume that one species is a herbivore (a pest, for our purposes) and its population in a field consists of 1 million individuals, while the second species is a predator of the first, with a population of 100,000 individuals. If the field is thoroughly treated with the pesticide, killing all susceptible insects, 10 individuals of the pest species will survive, but only I of the predator species will survive, each of the survivors being naturally resistant mutants. In this oversimplified example, the consequences are obvious. The small group of resistant pest insects can quickly reproduce a large population, free from attack by its predator.

Chlorinated Hydrocarbons

Chlorinated hydrocarbons include DDT, benzene hexachloride (BHC), dieldrin, endrin, aldrin, chlordane, lindane, isodrin, toxophene, and similar compounds designed to kill insects. DDT is the most thoroughly studied of the chlorinated hydrocarbons, and much of the following discussion is based on what is known about it. In its behaviour it is more or less typical of the group, although other chlorinated hydrocarbons may be more soluble in water, more toxic, less persistent, and so forth.

In insects and other animals these compounds act primarily on the central nervous system in ways that are not well understood, but the effects range from hyperexcitability to death following convulsions and paralysis. There is evidence that DDT interferes with the movements of ions through the membranes of nerve cells—movements that must occur for nerve impulses to be propagated.

Effects of DDT outside the nervous system also seem linked to processes involving cell membranes, as, for instance, in the membranous energy-mobilizing apparatus of the mitochondria (the power plants of cells). Chronic effects on vertebrates include infiltration of fat into heart muscle and fatty degeneration of the liver, which is often fatal. Fishes and other aquatic animals seem to be especially sensitive to chlorinated hydrocarbons.

Their oxygen uptake is somehow blocked at the gills, causing death from suffocation. That chlorinated hydrocarbons apparently can influence the production of enzymes may account for their wide range of effects Chlorinated hydrocarbons tend to be highly soluble in fats and fatty tissues. In animals this means that they may be stored at sites remote from the primary active site in the nervous system and thus rendered relatively harmless.

These compounds are more toxic to insects than to mammals, primarily because they are absorbed more easily through insect cuticle than through mammalian skin. Chlorinated hydrocarbons vary greatly in their toxicity to plants.

Organophosphates

Organophosphates include parathion, EPN, malathion, Azodrin, diazinon, dichlorvos, TEPP, phosdrin, and numerous others. These poisons are descendants of the nerve gas Tabun (ethyl N-dimethylphosphoramidocyanidate), developed in Nazi Germany during World War II. The majority of them are cholinesterase inhibitors; they inactivate the enzyme responsible for breaking down a nerve "transmitter substance," acetylcholine.

The result in acute cases of poisoning is a blockade of synapses (connections between nerve cells) that use acetylcholine as a transmitter—often with lethal effect. Unlike chlorinated hydrocarbons, organophosphates tend to be unstable and nonpersistent; hence, they tend not to produce chronic effects in ecosystems or to accumulate in food chains.

They often, however, are extraordinarily toxic. For example, the LD_{50} (dose required to kill 50 per cent of a group of test animals) of TEPP when administered orally to white rats is only from 1 to a few milligrams per kilogram of body weight. It is so toxic it is not marketed today.

Organophosphates inhibit other enzymes as well as cholinesterase. Indeed, some of the ones that show relatively great toxicity for insects and little toxicity for mammals do so because they poison an esterase that is more critical to the functioning of insect than of mammalian nervous systems. Malathion, which is violently poisonous to insects, is relatively nontoxic to mammals (acute oral LD_{50} in rats 900-5800 mg/kg) because mammalian liver and, to a lesser degree, other tissues contain the enzyme carboxyesterase, which destroys malathion.

But relatively high toxicity to mammals can occur when malathion is used in combination with some other organophosphates (such as EPN) that apparently inhibit carboxyesterase. For example, in dogs one-fortieth of the LD_{50} of malathion administered with one-fiftieth of the LD_{50} of EPN killed *all* the experimental animals—a classic example of a synergistic effect.

While most organophosphate pesticides produce acute poisoning, with transient loss of muscular strength and coordination due to their inhibition of cholinesterase, a few of these compounds have a delayed neurotoxic effect. For example, one organophosphate, Phosvel (leptophos), has a very low acute oral toxicity in chickens (4700 mg/kg), but after receiving doses ranging from 180 to 3000 mg/kg, which did not cause immediate symptoms, individuals developed irreversible, lethal neurological problems.

The symptoms of this nerve poisoning appeared only after a delay of 8 to 13 days. Phosvel has been implicated in a kill of some 1300 water buffalo in the Nile delta in 1971, and employees of a Phosvel manufacturing plant have developed delayed neurological symptoms.

Carbamates

Insecticides of the carbamate group include Sevin (carbaryl), zectran, baygon, dimetilan, pyrolan, and Temik (aldicarb). Chemically, these are esters of a derivative of carbamic acid. Attention originally was called to this group of compounds in the nineteenth century because one of them, eserine (physostigmine), proved to be the active ingredient in a poisonous plant, *Physostigma venenosum,* used in trials by ordeal by West Africans.

Synthetic analogues of eserine were used in medicine, but they had no insecticidal properties. Synthetics with insecticidal properties were first developed by Geigy Company in Switzerland in 1947.

Like organophosphates, carbamates inhibit cholinesterase, and some have very high acute toxicity (for example, aldicarb is in the same class as TEPP) They were introduced in many instances as substitutes for chlorinated hydrocarbons when insects developed resistance to the latter. They are quite selective in their action, as some insects are highly susceptible to these poisons and others are quite resistant. They also show variation in their degree of persistence in the environment, but generally they are much less persistent than chlorinated hydrocarbons.

If dosage of the insecticide is increased, the pest species will respond bv becoming more and more resistant with each generation. Not only herbivorous pests but other large populations can easily evolve such resistance; for example, resistance to DDT has developed in many mosquito populations, which has hampered malaria-control programs.

Artificial poisons tend to be much more effective against predators and parasites than against herbivores for another reason—the long coevolutionary experience herbivores have with plant poisons. Herbivores that were not adept at circumventing poisons are mostly long since extinct.

Small wonder that the evolutionary "winners" now being assaulted by mankind's poisons have so often managed to evolve resistance to them rapidly. Indeed, applications of pesticides often lead to *higher* concentrations of pests rather than lower, either because they directly poison the natural enemies of the pest or because they indirectly decrease the efficiency of those enemies as control agents. They accomplish the latter in several ways.

One is through the very success of the pesticide in reducing the target pest population. This may deprive the pest's natural enemies of their prey and thus reduce the predator or parasite population to a level where it is ineffective as a control when the pest population resurges. Similarly, the natural enemies of the pest may be decimated if the pesticide kills off another species preyed on by the same predator or parasite. Such an alternate host may be necessary to keep the predator population going all year round; if it is destroyed, the natural control is destroyed also.

There are, of course, many actual cases in which differential kill of predators has released some prey species from their natural restraints. It is fair to say, for instance, that mites, as pests, are a creation of the pesticide industry. Careless overuse of DDT and other pesticides has promoted many of these little insect like relatives of spiders to pest status by killing the insects that previously kept them under control.

The emergence of the European red mite as a major pest in apple orchards followed the use of DDT to control the codling moth. This is only one of the many examples in which pesticides intended for one pest have led others to flourish. A common response by the pesticide industry to such situations is to recommend the use of heavier doses of the original pesticide or to develop more potent poisons, which then create another array of pollution problems.

In such cases one or more pests were being held in check by natural biological controls, which were upset by the introduction of the chemical. That so many organisms can be "promoted" to pest status by the killing of predators and parasites with pesticides is itself a powerful testimonial for reliance on biological controls, instead. Such controls are constantly at work suppressing potential pests at no economic cost.

EFFECTS OF INSECTICIDES ON NONTARGET ORGANISMS

Of all the synthetic organic pesticides, probably more is known about DDT than any other. It is the oldest and most widely used chlorinated hydrocarbon insecticide. Although there has been an almost total ban on the use of DDT in the United States since 1972 and its use has been prohibited in some other countries, worldwide use in the mid-1970s was probably as great as in the mid-1960s.

DDT is found everywhere—not only where it has been applied, but all over the Earth. residues in the fat deposits of Eskimos and in those of Antarctic penguins and seals. Seals from the east coast of Scotland have been found to have concentrations of DDT as great as 23 ppm in their blubber, and both DDT and DDE contaminate fishes living 2500 meters deep in the ocean. Pesticide pollution is truly a worldwide problem.

The Impact of Chlorinated Hydrocarbons on Birds

Because DDT breaks down so slowly, it lasts for decades in soils. For instance, in the Long Island estuary studied by Woodwell and his colleagues the marsh had been sprayed for twenty years for mosquito control. Up to 32 pounds per acre of DDT were found in the upper layer of mud there. Such concentrations in U.S. soils are not unusual. As a result of the progressive concentration of DDT as it moves up food chains, the danger to the lives and reproductive capacities of fish-eating birds became extreme in the 1960s.

Nesting failures among bald eagles attributable to DDT reached proportions that brought the survival of the species into severe jeopardy. In addition, reproductive difficulties in populations of such diverse birds as the peregrine falcon and the brown pelican have been traced to residues of DDT. DDE, a breakdown product of DDT, interferes with the birds' ability to transport calcium, which results in their laying eggs whose shells are so thin that they are crushed by the weight of the incubating parents.

Similar effects have been caused experimentally by polychlorinated biphenyls (PCBs), which are chlorinated hydrocarbons used extensively in industry and which are closely related to the insecticides, as well as by mercury compounds. If they reach high enough levels in the environment they could cause problems for bird populations.

The evidence against DDT in the case of birds' eggs is now overwhelming. Studies have been made of the thickness of the eggshells of raptorial birds in museum collections. Among the eggs of those species investigated that feed high on food chains, there was in virtually all cases a sharp decrease in the thickness of shells between 1945 and 1947, when DDT was generally introduced. DDT has been directly implicated in eggshell thinning as early as 1948.

In the United Kingdom a strong correlation has also been shown between the level of chlorinated-hydrocarbon contamination in various geographical regions and the thickness of eggshells of peregrine falcons, sparrow hawks, and golden eagles nesting in those regions. The exceptions make the case even more convincing. The shells of one Florida population of bald eagles first thinned in 1943, not between 1945 and 1947.

Investigations disclosed that population lived in a county where large-scale DDT testing took place in 1943. One pair of peregrines on the coast of California was an exception to the general rule of nesting failure. Examination of their nests revealed that they were feeding inland on mourning doves. Mourning doves are herbivores; therefore the peregrines were feeding much lower in the food chain than if they had been feeding on fish-eating seabirds. They picked up less DDT and were therefore able to reproduce successfully.

Experimental feedings of DDT to mallard ducks, American sparrow hawks, and Japanese quail all produced thin eggshells and reduced hatch rates. DDE, a widely distributed, stable breakdown product of DDT, has also been shown to reduce the thickness of mallard eggshells and to reduce hatchability. Sensitivity to DDE varies greatly from species to species. Brown pelicans, cormorants, and raptors show high sensitivity; ducks, gulls, and ringdoves, moderate sensitivity; and quail, chickens, and Bengalese finches, low sensitivity.

At one time it was thought that the thin eggshell phenomenon was produced by alterations in estrogen levels, which are involved in the storage of calcium in birds. However, biologist D. B. Peakall of Cornell University did a series of experiments in which DDE was injected into ringdoves shortly before they laid eggs. The eggs, when laid, had thin shells. The DDE greatly reduced the activity of carbonic anhydrase, an enzyme that plays a critical role in providing calcium for eggshell production. It is now thought that inhibition of that enzyme and ATPases involved in transporting calcium from blood to eggshell is the basic mechanism that produces the thinning.

Dieldrin injected shortly before eggs are laid does not cause thinning of the eggshells. It and other chlorinated hydrocarbons (especially PCBS) do, however, lower the estrogen level, which in turn delays breeding. DDT does not have as great an estrogen effect as dieldrin, nor is it apparently as potent an inducer of liver enzymes.

Late breeding is an important component of observed declines in the numbers of predatory birds, since breeding is normally timed to coincide with optimal food supplies for the young. Interestingly, some of the most spectacular declines in the populations of these birds occurred years after thin eggshells first appeared-about the time that dieldrin was introduced.

There was, fortunately, a steady decrease in the amount of DDT used in the United States during the 1960s, use at the end of the decade being on the order of 25 per cent what it was at the start. At the end of 1972 its use was almost totally banned. Few reductions in DDT (or other organochlorine pesticides) residues were reported before 1970, but by 1974 a steady decline in the concentrations of DDT and its breakdown products DDD and DDE had been detected in migratory songbirds and in the eggs of brown pelicans.

In the United Kingdom, where use of organochlorines had been restricted, investigators were able to show an increase in the breeding success of Scottish golden eagles that was associated with a decline in the DDE and dieldrin content of their eggs, and they also recorded a decrease in the organochlorine content of cormorant (shag) eggs.

Shell-thinning is not the only way in which DDT can affect birds. For instance, it can also influence behaviour associated with migration. Such effects should decrease if the use of this dangerous chemical were phased out. Thus, if the use of organochlorines declines, there seems to be every reason for optimism that the effects of those poisons on the ecosystem will abate.

The impact of chlorinated hydrocarbons on aquatic organisms. Birds are not the only organisms that have suffered from chlorinated hydrocarbons. Coho salmon in Lake Michigan, which also feed at high trophic levels, have been laying eggs that contain DDT residues. In 1968 almost 700,000 young salmon died as they absorbed the last DDT-rich drop of oil from their yolk sacs. Trout have had similar problems, and DDT has been found in such commercially important marine fishes as tuna, mackerel, and hake.

One spectacular example of the destruction of nontarget organisms by pesticides was described in detail by reporter Frank Graham, Jr., who recounted the story of a great Mississippi fish kill in the early 1960s. Rough estimates gave the total loss in the four years from 1960 to 1963 as between 10 million and 15 million fishes in the lower Mississippi and its bypass, the Atchafalaya. The fishes killed included several kinds of catfish, menhaden, mullet, sea trout, drumfish, shad, and buffalo fish.

The die-offs were ruinous to the local fishing industry. A thorough investigation by government and private laboratories placed the blame primarily on the highly toxic chlorinated hydrocarbon insecticide endrin and one of its derivatives. They were found not only in the blood and tissues of dying fishes and water birds but also in the mud in areas where fishes were dying. In experiments, extracts made both from the mud and from the tissues of dying fishes killed healthy fishes.

It was found that fish kills were greatest in 1960 and 1963, when endrin was used commonly to treat cotton and cane crops in the lower Mississippi valley, and that fish kills were smallest in 1961 and 1962, when very little endrin was used.

The Public Health Service found that runoff from agricultural lands following dusting and spraying was one major source of endrin. The other source was the Memphis plant of the Velsicol Chemical Corporation; waste endrin from the manufacturing process was getting into the river. Graham described the reaction of the Velsicol Corporation:

"Velsicol, under fire, shot back. Bernard Lorant, the company's vice-president in charge of research, issued strong denials. In a statement to the press, he said that endrin had nothing to do with the Mississippi fish kill, that the symptoms of the dying fish were not those of endrin poisoning, and that Velsicol's test proved that the fish had died of dropsy." Unfortunately for the Velsicol defence, dropsy happens to be a disease that is never *epidemic* in fishes.

At a May 1964 conference in New Orleans on the Mississippi fish kill, a theme promoted by the pesticide industry was that only communist sympathizers would criticize the extravagant way pesticides are used. Their approach was exemplified by a statement made well before the conference by Parke C.

Brinkley, then president of the National Agricultural Chemicals Association: "Two of the biggest battles in this war [against communism] are the battle against starvation and the battle against disease. No two things make people more ripe for Communism. The most effective tool in the hands of the farmer and in the hands of the public health official as they fight these battles is pesticides."

This kind of response is found all too often in industries whose products are found to be dangerous. Finding ways to avoid the introduction and wide deployment of such products is a great social challenge. Once a great deal of money is committed to them, their producers tend to defend them vigorously, regardless of the risks to themselves or society.

In June 1969 a vast fish kill occurred in the Rhine River. Preliminary investigations indicated that containers holding at least 90 kilograms of the chlorinated hydrocarbon insecticide Endosulfan, which had dropped off a barge, were responsible. The increase in public understanding between 1964 and 1969 of the ecological dangers of pesticides was perhaps indicated by the refreshing absence on that occasion of comments from the pesticide industry about either dropsy or Communists.

Some of the most devastating effects of pesticides in aquatic systems come from releases in manufacture and transport. Just as the Velsicol plant appears to have been a major source of endrin in the Mississippi, so wastes from the Montrose Chemical plant in Los Angeles County, the sole producer of DDT in the United States, appear to have been the major source of residual DDT concentrations in organisms in the coastal waters of California in 1969-70.

Sand crabs near the sewage outfall carrying the Montrose discharge showed DDT concentrations over 45 times as great as they did near major agricultural drainage areas. This is encouraging, since measures to prevent accidental releases from factories are much more readily devised than measures to prevent the movement of pesticides from the fields where they are applied.

Another example will illustrate how complex, subtle, and far-reaching the effects of pesticide pollution may be. Ecologist L. B. Slobodkin, of the State University of New York, has described a plan to block the seaward ends of lochs in western Scotland and use them as ponds for raising fishes.

One of the problems was to find ways to raise the young fishes in the laboratory so they could be "planted" in the ponds. It was discovered that newly hatched brine shrimp would serve as a satisfactory food for the kinds of fishes that would be raised. The shrimp could be obtained from brine-shrimp eggs gathered commercially in the United States and sold to tropical fish fanciers for feeding young tropical fishes.

The supplies came from two places: San Francisco Bay and the Great Salt Lake. But sufficient eggs for the project could no longer be obtained from San Francisco Bay because of the demands of local aquarists and because large areas of former brine shrimp habitat had been transformed into residential subdivisions.

Unfortunately, the Utah supply was of no use to the project either, because brine shrimp hatched from Utah eggs killed the young fishes in the United Kingdom. The Utah shrimp had absorbed the insecticides that drained into the Great Salt Lake from surrounding farmlands. Thus, insecticide pollution in Utah hampered fish production in Scotland!

The list of nontarget animals affected by insecticides is now of encyclopedic proportions. The amazing diversity of organisms involved, besides many beneficial insects, includes robins, amphipods, planarians, whitefish, earthworms, old-squaw ducks, warblers, rabbits, quail, lake trout, mosquito fish, bluegills, foxes, oppossums, mice, ospreys, muskrats, pheasants, turkeys, gulls, fiddler crabs, salmon, frogs, toads, snakes, big-game mammals, and various plants.

Poisoning of these or any other organisms, of course, always affects the ecosystems in which they function.

Chapter 12

Insecticides and Marine Food Chain

The effects of insecticides in water are not confined to aquatic animals. Perhaps the most disturbing ecological news of 1968 was contained in the short paper "DDT reduces photosynthesis by marine phytoplankton." The author, environmental scientist Charles F. Wurster of the State University of New York, reported that DDT reduced photosynthesis in both experimental cultures and natural communities of marine phytoplankton (such as algae and diatoms), the tiny green plants that float free in the waters of the oceans.

Effects were noted at DDT concentrations of only a few parts per billion, quantities commonly suspended in waters near land sites treated with DDT. Water at a distance from treatment sites, however, ordinarily has DDT concentrations averaging less than one part per billion.

Similar results have been confirmed for DDT, dieldrin, and the related chlorinated hydrocarbon pollutants, PCBS. Dieldrin has been shown not only to inhibit growth in dinoflagellates but to cause cell disintegration at concentrations of 100 ppb. Moreover, the size reductions persisted for several generations after removal from dieldrin exposure.

The vulnerability of phytoplankters to chlorinated hydrocarbons varies with environmental factors such as temperature, light intensity, and interspecific competition, as well as with geographical location. Recent work has indicated that it is not the photosynthetic process of the phytoplankton that is affected by the chlorinated hydrocarbons but the rate of division of the plant cells.

The possible effects of DDT on phytoplankton in nature are difficult to evaluate. Phytoplankton are the primary producers responsible for most of the food taken from the sea. If photosynthesis by marine phytoplankton were significantly reduced, the amount of life in the seas would be reduced; if marine photosynthesis ceased, virtually all sea life would die.

But there is no sign that chlorinated hydrocarbons would ever reach a level in the marine environment that would pose any threat to the phytoplankton community as a whole. In response to chlorinated-hydrocarbon pollution, significant qualitative changes in the phytoplankton community seem more probable than large quantitative changes.

Phytoplankton populations are differentially susceptible to chlorinated hydrocarbons; even extremely low concentrations might result in shifts of dominance, leading to huge blooms of one or a few species. Indeed, changes in dominance relations have been recorded at chlorinated-hydrocarbon concentrations that had no detectable effect on single species cultures. Such shifts could, in turn, produce serious consequences throughout the oceanic food webs.

One possibility is that phytoplankton less acceptable to large herbivores would dominate, so that larger animals would be affected. This shortening of food chains might leave only microscopic plants and animals as sources of food from the sea. Another possibility is that phytoplankton communities near shore could become dominated by smaller species, thereby lengthening food chains and dramatically reducing the sizes of populations of fishes at the upper trophic levels.

All of this is, of course, purely speculative, especially since levels of pollution found to cause shifts in photosynthetic rates have so far been found only extremely rarely in marine environments. As use of these pollutants declines, this once seemingly serious threat to the oceans may prove to be a non problem.

Similarly, the problems besetting inland waters could be aggravated by the effects of DDT on freshwater phytoplankton. On the other hand, DDT or other insecticides may actually cause phytoplankton blooms because zooplankters, which normally limit the phytoplankters by grazing on them, may be even more susceptible to the pesticides. As C. F. Wurster wrote, "Such effects are insidious and their cause may be obscure, yet they may be ecologically more important than the obvious, direct mortality of larger organisms that is so often reported."

Insecticides and Soils

The effects of pollutants on soils are difficult to evaluate. Soils are not just collections of crushed rock; they are extraordinarily complex ecosystems. The animals of the soil are extremely numerous and varied. In forest communities of North Carolina, for example, an estimated 300 million small invertebrates live in each hectare of soil, about 30,000 per square meter.

Some 70 per cent of them are mites, a group of arthropods that may eventually be found to be as diverse as insects. In a study of pasture soils in Denmark, as many as 45,000 small oligochaete worms, 10 million nematodes (roundworms), and 48,000 small arthropods (insects and mites) were found in each square meter.

Even more abundant are the microflora of the soil. More than a million bacteria of one type may be found in a gram (0.035 ounce) of forest soil, as well as almost 100,000 yeast cells and about 50,000 bits of fungus mycelium. A gram of fertile agricultural soil has yielded over 2.5 billion bacteria, 400,000 fungi, 50,000 algae, and 30,000 protozoa.

The beneficial effects of earthworms are well known, but most people are unaware of the myriad other complex (and in many cases still poorly understood) ecological relationships within the soil that make it suitable for the growth of oak trees, sagebrush, corn, or any other plants. The soil contains microorganisms that are responsible for the conversion of nitrogen, phosphorus, and sulfur to forms available to the plants.

And many trees have been found to depend on associations with fungi. The fungi get carbohydrates and other essential substances from the roots, and the root fungus complex is able to extract from the soil minerals that could not be extracted by the root alone.

Such mycorrhizal associations are just beginning to be understood, but it is clear that in many areas the visible plant community would be drastically altered if the mycorrhizal fungi were absent from the soil.

Recognizing, as they do, that most of the complex physical and chemical processes responsible for soil fertility are dependent upon soil organisms, environmental biologists are opposed to the continuing treatment of soils with heavy dosages of deadly and persistent poisons. Consider, for instance, a study of the persistence of ten different chlorinated hydrocarbons in a sandy loam soil at an experimental station.

Half of these insecticides had more than one-third of the amounts applied remaining in the soil fourteen or more years after treatment. Under normal agricultural conditions, chlorinated hydrocarbons.

Considerable evidence already exists that the use of insecticides may reduce soil fertility, especially in woodland soils, which are subject to spraying but not artificial cultivation. Populations of earthworms, soil mites, and insects are dramatically changed, and they in turn affect the soil fungi, which are their principal food. Even if bacteria were not affected directly, there is no question that the general effects on the soil ecosystem would carry over to them and to other microorganisms.

But it would be unwarranted to assume that the bacteria were not directly affected. It is known that a few microorganisms can degrade DDT to DDD under the proper conditions; a few can degrade dieldrin to aldrin and several other breakdown products of unknown toxicity.

Table. Persistence of Insecticides in Soils

Insecticide	Years since treatment	Percent remaining
Aldrin	14	40
Chlordane	14	40
Endrin	14	41
Heptachlor	14	16
Dilan	14	23
Isodrin	14	15
Benzene hexachloride	14	10
Toxaphene	14	45
Dieldrin	15	31
DDT	17	39

Our ignorance of the interactions of insecticides, herbicides and other environmental poisons with soil microorganisms is immense.

Agriculture, Ecology, and Pest Control

Pest-control, especially insect-pest control, in agriculture today is an ecological disaster. Agronomists understandably breed out of plants some of their natural chemical defenses against the pests. Various herbivore poisons also poison people and most taste unpleasant, although some spices, which are used in small quantities, are produced by plants as pesticides.

Farmers plant crops in tight, simple monocultures, inviting pest outbreaks, to which they then respond with synthetic pesticides, often killing a higher proportion of some nontarget populations than of the target population of pests.

Because synthetic pesticides, whether intended for pest animals or pest plants, have toxic effects on so many nontarget organisms, they are more accurately called *biocides*. And, because those nontarget populations often include predators and parasites of pests or potential pests, the pesticides often make pest problems worse.

Pesticides and Crop Losses

There are a few signs that ecologically sound agricultural practices may eventually be adopted, but so far the general trend has been in the opposite direction. The percentage of crop losses to insects in the United States seems not to have been reduced since 1945, despite the enormously increased use of powerful pesticides. In 1948, William Vogt noted in his book *Road to survival* that "one-tenth of all crop plants are destroyed by insects in the U.S. every year."

Vogt based his comment on statistics published by the U.S. Department of Agriculture. Georg Borgstrom, also using Department of Agriculture figures, observed that crop losses due to insects in the 1960s amounted to the yield from about one-fifth of the total acreage, or about one-sixth of the cash value of the total crop production in the United States When the storage-loss component was subtracted, the field losses amounted to slightly more than one-tenth of total production — about the same as 1948.

The President's Science Advisory Committee panel on the World Food Problem estimated insect losses in the field during the 1950s as between 4 and 14 per cent, depending on the crop.

In 1948, according to zoologist Robert L. Rudd of the University of California, DDT, benzene hexachloride, and lead arsenate were the only insecticides of any significance used. By 1958, production of DDT and other chlorinated hydrocarbon insecticides (some of which was exported) in the United States amounted to nearly 12 times the 1948 production of DDT. Overall, pesticide usage in the United States more than doubled between 1960 and 1970, with about 500 million kilograms being used in 1972.

Rudd pointed out that it is difficult to get accurate estimates of pest losses and that standards of pest damage have changed through time. Nevertheless, the consistency of estimates of losses caused by pests between the 1940s and 1960s was rather striking. Indeed, since 1965 the percentages of many major crops lost to pests may have been *rising*. Thus, despite huge inputs of biocides, insects and diseases have been claiming a constant or growing share of the American farmers' greatly increased agricultural production.

What proportion of that increased production can be attributed to the use of synthetic pesticides? Certainly not as much as the pesticide industry would like us to believe, but synthetic pesticides may play a more significant role than is indicated in those rather dismal percentage-loss figures. High-yield strains of crops, heavily fertilized, probably require more protection per acre than lower-yield strains. That is, one might expect even greater percentages of some crops to be lost to pests today if we were still using the control techniques of 1940.

But on others the use of synthetic pesticides has decreased crop yields. According to University of California entomologist Robert van den Bosch (one of the nation's outstanding insect ecologists and an expert on ecologically sound control measures), in some California cotton fields harvests *increase* when pesticides are withdrawn. Pest problems on that state's important citrus crops are now about 3 times as severe as they were in the 1940s.

Ten per cent of the world's insecticide load is applied to 5.7 million hectares of irrigated California land, and the state is rapidly becoming an insecticide disaster area. Of the 25 most serious insect pests in the state (each causing more than $1 million in damage), about 85 per cent are resistant to one or more insecticides and all are involved in pesticide-aggravated problems -that is, they commonly have resurged following pesticide application, or they were not pests until insecticides changed the ecological situation, or pesticides applied to control them engendered outbreaks of other pests.

Uncertainties about losses relative to the intensity of today's control practices make it difficult to evaluate the success of these practices, even ignoring the key questions of the undesirable side effects of pesticides or whether alternative strategies adopted decades ago would have provided superior control. Could, for instance, similar or greater yields have been achieved since World War II with control methods that were more ecologically sophisticated than current pesticide practices?

It is the opinion of systems ecologist K. E. F. Watt of the University of California that, when measured against successful biological and integrated control programs, "most pesticide projects have been failures." It is worth noting here that the Chinese have independently arrived at the same conclusion and have been actively developing other modes of pest control.

We are convinced that the pest-control procedures in use in the 1950s and 1960s will eventually be seen as one of humanity's most tragic blunders and that when the total accounting is done, it will reveal that other methods of control would have provided higher yields at less direct cost and with fewer deleterious consequences for human beings.

Case Histories of Insecticide Failures

Pesticides, especially the persistent ones, simplify the ecosystems to which they are applied and those into which they are transported by wind and water. The results of direct application are all too often the same: pests are freed from natural restraints. A pest, at first seemingly controlled, is soon back in even larger numbers than before.

This destabilizing effect is so common that it is often cited in the scientific literature as evidence that ecosystem simplification leads to instability. Unfortunately, things do not end there in agricultural practice, and higher doses of pesticides are applied, thus aggravating the situation still further.

The history of attempts to control cotton pests in the coastal Capete Valley of Peru has been reported by entomologist Ray F. Smith of the University of California. Against the advice of entomologists who recommended the use of cultural-control methods (such as destroying weeds which harbor pests) and inorganic and botanical insecticides, synthetic organic pesticides were widely introduced in that valley in 1949. At first the use of the pesticides, principally the chlorinated hydrocarbons DDT, BHC, and toxaphene, was very successful.

Cotton yields increased from 494 kilograms per hectare (440 lbs/ acre) in 1950 to 728 kilograms per hectare (648 lbs/acre) in 1954. The cotton farmers concluded that if more pesticide were applied, more cotton would grow.

Insecticides "were applied like a blanket over the entire valley. Trees were cut down to make it easier for the airplanes to treat the fields. The birds that nested in these trees disappeared. Other beneficial animal forms, such as insect parasites and predators, disappeared. As the years went by, the number of treatments was increased; also, each year the treatments were started earlier because of the earlier attacks of the pests."

Trouble started in 1952, when BHC proved no longer to be effective against aphids. In 1954 toxaphene failed against the tobacco leaf worm. Boll weevil infestation reached extremely high levels in 1955/ 1956, and at least six new pests had appeared, pests that were not found in similar nearby valleys that had not been sprayed with organic pesticides. In addition, the numbers of an old pest, larvae of the moth *Heliothis virescens*, exploded to new heights, and showed a high level of DDT-resistance.

Synthetic organic phosphates were substituted for the chlorinated hydrocarbons, and the interval between treatments was shortened from one or two weeks down to three days. In 1955/ 1956, cotton yields dropped to 332 kilograms per hectare, despite the tremendous amounts of insecticide applied. Economic disaster overtook the valley. In 1957 an ecologically rational integrated control programme was initiated, in which biological, cultural, and chemical controls were combined. Conditions improved immensely, and yields rose to new highs.

This example should not be taken to mean that ecologically unsound pesticide programs have been instituted only in other places and in former times. Many examples can be cited of mistakes made here in the United States. For example, in both California and Arizona, profit margins have been dropping for cotton farmers because of insect attacks and the rising costs of chemical pesticide applications. In an attempt to evaluate the control measures used by agriculturalists, biologists uncovered some astounding information.

The programs designed to control cotton pests in those states were reviewed by Kevin Shea, scientific director of the U.S.-based Committee for Environmental Information. Shea reported, "controlled experiments on small plots suggest that cotton growers in California may have spent thousands of dollars to fight an insect, the lygus bug, that has no appreciable effect on the final production per acre." Reduction of the number of lygus bugs per acre *did not lead to increased yields*, apparently because the bugs fed on "surplus" squares (flower buds) of cotton, which would not ripen in any case.

Spraying for lygus bugs in mid-season with some pesticides not only did not help yields, but lowered them by killing the insect predators of the bollworm, which caused outbreaks of bollworms and other pest caterpillars. In fact, by the mid-1970s the bollworm had virtually disappeared as a pest of cotton in the San Joaquin Valley. This had occurred because of adjustments in the lygus spray programme (that is, sprays have been limited to restricted acreages in the early part of the season).

Most disturbing of all is the Azodrin story. Azodrin is a broad-spectrum organophosphate insecticide manufactured by the Shell Chemical Company (a subsidiary of the Shell Oil Company). Azodrin kills most of the insect populations in a field, but like other organophosphates (and unlike the cholorinated hydrocarbons) it is not persistent. Its effects are devastating to populations of predatory insects.

Therefore, when a field is reinvaded by pests, or when pest survivors make a comeback, their natural enemies are often absent, and pest-population booms may occur. Experiments by van den Bosch and his colleagues clearly indicated that, rather than controlling bollworms, Azodrin applications, through their effect on the bollworms' natural enemies, actually *increased* bollworm populations in treated fields.

Other Azodrin experiments, and the similar results obtained with the use of other broad-spectrum pesticides, make it clear that the control procedure, rather than helping the farmer, often has precisely the opposite effect. One might reasonably expect the pesticide manufacturer to withdraw the product, or at least warn customers and advise them how to avoid the disastrous effects. However, Shea wrote:"

Shell Chemical Company, manufacturer of Azodrin, was aware of the University's findings through publications and from a seminar given at the Shell Research Centre in Modesto. Nevertheless, the Company decided to promote the material for use on cotton pests in the San Joaquin Valley. Shell mounted a massive sales campaign. Radio, television, billboards, bumper stickers, and trade journal ads were employed to move Azodrin out of the warehouses and into the fields.

Azodrin was heralded as being unmatched in its ability to kill every major insect likely to damage cotton." The Modesto seminar was given by van den Bosch, who has said, "our own experiments indicate considerable odds against significant economic gains resulting from the use of Azodrin."

Shell further has been promoting the use of Azodrin on a fixed schedule, whether pests are present or not. But fixed-schedule spraying unnecessarily damages nontarget organisms, destabilizes the agricultural ecosystem, and creates pest problems that otherwise would not arise.

Aside from enhancing those effects, such programs, of course, promote the development of resistant strains of pests and the destruction of natural enemies. This guarantees the "need" for heavier and heavier doses of Azodrin. As one Azodrin advertisement put it, "even if an overpowering migration (*sic*) develops, the flexibility of Azodrin lets us regain control fast. Just increase the dosage according to label recommendations."

Pesticides and the Department of Agriculture

It would be unfair to blame the petrochemical industry alone for the misuse of pesticides. The U.S. Department of Agriculture (USDA) has also contributed heavily to environmental deterioration resulting from such misuse. That agency has long uncritically promoted pesticides, often displaying a distressing degree of ignorance of ecology in the process.

An outstanding example can be found in the history of the fire-ant programme, in which the USDA in the late 1950s attempted to "exterminate" that insect over a large portion of the United States by spraying with the chlorinated hydrocarbons dieldrin and heptachlor. Many independent biologists protested the programme, asserting that extermination of the ants would be virtually impossible and that the unwanted side effects would be horrendous. Undaunted, the USDA forged ahead, but by 1960 it was clear that the programme was a fiasco, as predicted.

With respect to the fire ant, however, the USDA apparently was incapable of learning. Armed with a new chlorinated-hydrocarbon insecticide, mirex, the department announced plans in 1970 to spray bait pellets over some 4.5 million hectares in the South. During the previous eight years, mirex had already been used over large areas on an experimental basis against the fire ant.

It was at first touted as the perfect pesticide for the programme, because virtually all the bait found its way into the ants' nests, leaving very little to poison nontarget organisms, and it appeared to be generally less toxic than other chlorinated hydrocarbons such as dieldrin or heptachlor. But, as time and research went on, it was discovered that mirex was not so harmless, after all. Like other chlorinated hydrocarbons, it is persistent and tends to be concentrated in food chains.

In laboratory tests, it has been found to be highly toxic to shrimp, crabs, and crayfish, as well as to insects other than the fire ant. It is also somewhat toxic to at least some birds and mammals and is carcinogenic in mice. Because of over treatment (repeated applications after most of the ant nests had been poisoned), enough mirex escaped into the environment to cause problems.

When the USDA announced the new programme, the Environmental Defence Fund, together with several other conservation groups, filed suit against it. In April 1971, their motion for a preliminary injunction against the fire-ant programme was denied, and the programme was initiated immediately. However, the action had served to delay its start. Moreover, perhaps partly because of the suit and partly because some states failed to produce their shares of the funds for it, the originally planned programme was considerably curtailed.

A number of other government agencies became interested in mirex. The Department of the Interior put mirex on its list of restricted pesticides, and the Department of Health, Education and Welfare disapproved of mirex because of its carcinogenic properties. The Council on Environmental Quality also recommended against the use of mirex, and the Environmental Protection Agency (EPA) in 1972 limited mirex applications to once a year and to areas away from forests and waterways (three times a year without spatial restriction had been the practice).

In 1975 the USDA at last abandoned the goal of eradicating the fire ant. Secretary of Agriculture Earl L. Butz gave as the reason the EPA restrictions, which he said would turn USDA operations "into a control programme of living with the ants rather than working toward an eradication programme to wipe them out." In late 1976, the USDA and the EPA settled the six-year-old battle over mirex with an 18-month phaseout programme.

But it is likely that the USDA will continue the attack on fire ants with some new compound. Although *no* widespread insect pest has ever been eradicated by pesticides, and none is ever likely to be, this news has yet to penetrate the appropriate recesses of the USDA. In 1974, for instance, the department seriously recommended a scheme to "eradicate" the boll weevil. Whether or not that programme will be attempted remains to be seen Whether it would be successful is less in doubt.

Not all pest-control programs initiated by the USDA have been so incompetent. Some ecologically sophisticated programs have been initiated within the department. Perhaps the most brilliant was that against the screwworm, a fly whose larvae (maggots) can be an extremely serious pest on cattle. Annual losses in livestock because of that pest were estimated to have been as much as $40 million a year in the early 1960s.

Under the leadership of entomologist E. F. Knipling, the USDA embarked on a massive programme of sterilizing male screwworm flies by irradiation and then releasing them in infested areas. The female screwworm only mates once. By flooding infested areas with sterile males, the screwworm was essentially eradicated from the southeastern United States. The effectiveness of that biological control programme makes an interesting contrast with the futile and destructive fire-ant programme.

ALTERNATIVE METHODS OF CONTROLLING POPULATIONS OF INSECTS AND MITES

There are three main reasons why the broadcast use of synthetic poisons is a generally undesirable way of controlling the sizes of pest populations. The second is that, by affecting nontarget organisms, synthetic poisons may disrupt the crucial service functions of ecosystems upon which human beings depend.

One of those functions is, ironically, the control of some 98 or 99 per cent of the potential pests. When that function is disrupted by the use of pesticides, the result, as in the Capete Valley, is the promotion of organisms that have been released from natural controls to pest status.

The third reason the broadcast use of synthetic poisons is undesirable is that such use encourages the development in pests of resistance to the pesticides. This problem is so serious that it has been described as a "primary obstacle to successful pest control today." These failings mean that not only does broadcast spraying represent both direct and indirect hazards to human beings, but that very often the spraying does not control the pest — and even exacerbates pest problems. Nevertheless, we could gain the impression from pesticide industry propaganda that only current patterns of chemical control stand between us and death from insect-borne disease or starvation. Nothing could be further from the truth. There are, in many cases, alternatives to broadcast use of pesticides that are more effective in controlling pests and that pose little or no threat to people or ecosystems.

Biological Control by Predators and Parasites

Perhaps the best known of those alternatives is biological control using predators and parasites. In essence, this is employing the same technique "used" by natural ecosystems to achieve near-perfect control of potential pests. When a pest problem arises, organisms that attack the pest are introduced, and nature takes its course.

The first (and still most famous) case of biological control is the saga of the cottony-cushion scale. This scale insect, *Icerya purchasi*, a native of Australia, was discovered in California in 1868 and by the mid-1880s was in the process of destroying the California citrus industry. A parasitic fly and a predatory beetle were imported from Australia, where those and other natural enemies kept *Icerya* from becoming a pest.

The importations occurred in 1888 and 1889 and achieved spectacular success. By the end of 1889 Icerya had ceased to be a problem in California. The beetle (the Australian ladybird or vedalia beetle) achieved the first great success, spreading rapidly, but it is clear that the fly would have been equally successful, and today it is the primary control factor in coastal areas.

The spraying decimated populations of the vedalia beetle and caused a catastrophic population explosion of the cottony-cushion scale, which had been demoted to a nonpest status for more than half a century. Severe economic damage was done in just a few years until the spray programs were discontinued and biological control reestablished.

There are many other biological-control success stories, but they are little known to the public. At small expense and effort a pest problem can thus be solved once and for all. The continuing, expensive effort — often with ever-increasing failure of control and harmful side effects — that characterizes many chemical-control programs, is absent. Only monitoring is required to make sure the biological control continues to work.

By importing predators or parasites some degree of permanent control has been achieved on some 130 insect pest species. About one-third of them have been completely controlled, one-third substantially controlled, and one-third partially controlled. Some of the insects have been the targets of control programs in more than one country — all told, there have been worldwide some 275 successful biological-control projects.

All in all, there appears to be about a 50-per cent chance of achieving significant control of an insect pest by importing natural enemies to attack it.

Nevertheless, of some 5000 recorded species of insect pests around the world, attempts have been made to find and import natural enemies of fewer than 5 per cent. But, of course, there is no profit for the petrochemical industry if pests are controlled by natural enemies (and relatively little profit even if chemical controls themselves work).

Another type of biological control with the potential for being as effective as the use of arthropod enemies of insects is the dispersing of bacterial, fungal, and viral parasites that also attack the insects. Although up to 1959 there were only two such microbial insecticides on the market, by 1975 nearly fifty had been given trade names and were in commercial or experimental use. It has been estimated that, for example, the successful use of the pathogen *Baculovirus heliothis* to control pink bollworms on cotton could reduce the use of synthetic pesticides by at least 8 million kilograms annually.

The importation of natural enemies is not without risk, it must be noted. For example, the introduction of mongooses into Jamaica to control rats led to an ecocatastrophe in which the native fauna of birds and small land animals was decimated. Today, however, importing agencies have established extremely careful protocols to screen out such "bad actors," doing careful pre-importation experiments in special fail-safe quarantine facilities.

BIOLOGICAL CONTROL BY GENETIC METHODS

The sterile-male technique has already been discussed in connection with the screwworm programme in the southeastern United States. That biological-control programs once successful may not be permanently successful is shown by the history of attempts to control screwworms along the border between the United States and Mexico.

Conditions there were not as favorable to the technique, and initial successes in the mid-1960s were followed by a resurgence of screwworms in the early 1970s — a resurgence made more serious because ranches then lacked the skilled cowboys, trained horses, and necessary facilities to reinstitute older methods of control.

One of the problems was that in the sustained programme

necessary to prevent rein festation by screwworms from Mexico (in contrast to the "quick fix" in the southeast) maintaining the genetic quality of the stocks used for rclease required genetic monitoring, which was not done properly.

Unfortunately, it appears that the selection pressures inadvertently applied in the laboratory produced a strain in which the males could not compete with wild males. This led to a collapse of control. The USDA is now attempting to remedy these problems. Among other things, it is contracting for extensive genetic research. The crucial point is that no pest-control programme, biological or chemical, can ever really be considered a *permanent* success — monitoring and backup plans in case of resurgence or rein festation should always follow any successful programme.

The success of the screwworm programme has led to intensive investigations of other methods of *genetic control* of pest populations. A great deal of attention has been paid, for instance, to the use of natural sterility produced by chromosomal rearrangements, and other techniques such as inducing sex-ratio distortion have been considered. So far, successes have been limited to pilot studies, but there is every reason to believe that there will be future successes of the kind achieved with the initial screwworm programme.

Biological Control by Cultural and Hormonal Methods

Another form of biological control is so-called *cultural control,* in which conditions in the crop ecosystem are modified to the detriment of the pests or to the advantage of their natural enemies. Weeds or crop remains that harbor pests can be destroyed, planting or harvesting dates modified, crops rotated, several crops planted together, and so on.

The genetic enhancement of plant defenses to produce strains resistant to pests, to aid crops in their coevolutionary races with the organisms that attack them. Research is also being carried out with a view to isolating and making use of plants' natural chemical defenses against insects and plant diseases. For instance, one group of chemicals seems to inhibit the juvenile hormones of insects, thus disrupting their development.

The use of attractants, especially of pheromones (hormones that operate outside the body — in this case, sex lures used by insects to attract mates), has had limited but important success in pest control. It has been especially effective against tephritid fruit flies, which often do major damage in orchards and melon fields.

An artificial sex pheromone, methyl eugenol, has achieved several local eradications of *Dacus dorsalis*, the oriental fruit fly, and eugenol is used as the lure in traps that are monitored to detect importations of that important pest.

Integrated Control

There obviously is no single panacea to replace the broadcast spraying of pesticides. What is required is a shift to ecological pest management — that is, to what is often called *integrated control*. Integrated control has as its goal the maintenance of potential pest populations below the levels at which they cause serious health hazards or economic damage.

It does not attempt to exterminate pests — which, as stated earlier, has never been accomplished by the broadcast spraying of pesticides. Integrated control involves using one or more techniques appropriate to the particular pest situation. Mosquitoes may be controlled by draining swamps in which larvae live, stocking lakes with mosquito-eating fishes, or applying hormonal insecticides, or perhaps directly applying toxic nonpersistent insecticides to any standing water that will not support fishes and cannot be drained.

Similarly, a crop could be protected by practices such as planting it in mixed cultures with other crops, destroying pest reservoirs adjacent to fields, introducing and encouraging appropriate predators and parasites, enhancing the natural defenses of crop populations, luring pests from fields with baits, and using nonpersistent insecticides.

These and other practices may be combined to achieve both high levels of desirable control and a minimum of damage to the ecosystems of the world. Sometimes, as in the successful screwworm programme, no chemical control will be necessary; at other times chemical methods may play major roles.

Integrated control programs have been enormously successful where they have been tried. An outstanding example is a programme in cotton in California. Robert van den Bosch and his co-authors wrote about the control of pests on cotton in California's Central Valley, "Existing pest control practices are both inefficient and expensive, and because of this they are directly contributing to the economic crisis. Accumulating evidence indicates that integrated control is much more efficient and less expensive, and the hard-pressed growers are beginning to realise this."

The transition away from the relatively simple chemical techniques will require planning and training, and will perhaps produce some temporary economic stress. One county agricultural extension service entomologist, faced with a successful integrated-control programme, said, "Your damn pest management programme is ruining my business. I can't get pilots who are willing to work for me when there is so little insecticide being used."

Such attitudes are, quite naturally, widespread in the insecticide industry, where profits depend on the avoidance of permanent solutions to pest problems — other than perpetual spraying.

Since banning the chlorinated hydrocarbons should not be delayed, more expensive methods of control may have to be used temporarily. But there is no reason the transition should have serious consequences for human health or nutrition.

The step would be necessary, however, even if such consequences *were* foreseen as serious, since continued use of persistent insecticides risks an unprecedented catastrophe for the entire planet. In fact, positive benefits would be immediate in many areas, where the development of DDT-resistant mosquitoes is already reducing the effectiveness of mosquito-control programs.

It is important to remember, above all, that avoiding unhappy side effects from any control programme, integrated or not, depends on intelligent surveillance and periodic reevaluation. *Any* tinkering with ecosystems may have unforeseen and deleterious consequences.

Chapter 13

Herbicides in Ecosystems

In recent years there has been an enormous upsurge in the use of herbicides as a substitute for farm machinery and labour in cultivating crops to increase productivity of range and pasture land; for keeping roadsides, railroad rights-of-way, and power line cuts free of shrubs; and as military defoliants during the Vietnam war.

The rate of increase in herbicide use has far outstripped that of synthetic insecticides, and annual sales of herbicides are now larger. Several kinds of herbicides are in wide use. Members of one group (2,4-D; 2,4,5-T; picloram; and so on) are similar to plant hormones and cause metabolic changes in the plant that lead to death or leaf drop. Members of another group (simazin, monuron, and the like) interfere with a critical process in photosynthesis, causing the plant to die from lack of energy.

The way others disrupt plant metabolism is less well understood. Although their direct toxicity to animals is low, herbicides have great impact on animal populations, since all animals depend on plants for food, at least indirectly. Furthermore, as a result of the coevolutionary interactions discussed earlier, most herbivorous animals are specialized to feed on one kind or on just a few kinds of plants.

Herbicides That Mimic Indoleacetic Acid

This group of herbicides includes 2,4-D, 2,4,5T, picloram, and several others. The compounds are chemically similar to the growth-regulating substance indoleacetic acid. That plant hormone (an auxin) controls such diverse activities as shoot growth, root growth, apical dominance, and phototropism. Herbicides in this hormonelike group function by causing uncontrolled growth and metabolism — the plant, in essence, grows itself to death.

Indoleacetic acid does not function as a growth substance in animals, and therefore it is not surprising that these herbicides have essentially no direct effect on animals. Differential toxicity to broad-leaved plants (in contrast to narrow-leaved grasses) is a function of the greater ease with which the compounds are absorbed by the broad-leaved plants.

Some of the herbicides, such as 2,4-D, are rapidly metabolized in woody plants, which usually localize the damage (they are defoliated but not killed). Other herbicides, such as picloram, remain active for long periods in trees and ecosystems.

Triazines and Substituted Ureas

These two classes of compounds include atrazine, simazine, fenuron, diuron, and monuron. Both kinds of compound block a critical step in photosynthesis, known as the Hill reaction. The plant, in essence, starves to death. Since animals do not photosynthesize, they are not directly affected by these compounds, with the exception of possible mutagenic effects from the triazines.

A wide variety of other synthetic compounds, including chlorinated aliphatic acids (dalapon, TCA), phenols (DNBP), acetamides (propanil) and thiocarbamates, can act as herbicides. Their modes of action are varied, and many are poorly understood.

For instance, the carbamate sulfoxides interfere with plant metabolism and growth and are often highly selective, but the precise mechanism for their action has not been discovered. They are, however, highly biodegradable — and the mechanism of their metabolic destruction is fairly clear.

In the light of these considerations, it is possible to evaluate the statements made by many United States government officials concerning the lack of danger to animals or the uncertainty of effects on animals of our defoliation activities in Vietnam.

Defoliation of tropical jungles inevitably leads to the local extinction of many populations of insects, birds, arboreal reptiles, and arboreal mammals. But, of course, *animals* in official statements can all too often be translated "elephants, tigers, and other large mammals." In temperate forests there is generally a less distinct canopy fauna, but changes in animal populations in response to large-scale defoliation would certainly be tremendous.

Little is known about the direct effects of herbicides on soil microorganisms. Research in Sweden indicates that herbicides destroy bacteria that are symbiotic with legumes, although the bacteria apparently are capable of developing some resistance.

Some herbicides, such as 2,4-D, are quickly degraded by bacteria and persist for only a few weeks or months; others (2,4,5-T and, presumably, picloram) are more persistent. Soil microorganisms (primarily bacteria and fungi) do not photosynthesize; they are consumers, not producers. Therefore, they probably will not be affected by herbicides that block photosynthetic processes, although there could, of course, be other metabolic effects. Herbicides that function as simulated plant hormones are unlikely to disturb the growth processes of the soil flora, since there is no evidence that the plant hormone simulated by those substances functions in bacteria or fungi. Those herbicides may, of course, have other physiological effects, since, as noted, some are metabolized by soil bacteria.

Ecocide in Indochina

The practice of genocide, unhappily, has a long history in human affairs. Ecocide (the deliberate destruction of ecosystems) is for practical reasons a rather recent development. Destruction of enemy crops has frequently occurred in warfare, and the near-extermination of the bison was a decisive factor in America's conquest of the Plains Indians (the slaughter was not a purposeful military strategy, but its effects were known at the time).

But, until recently, humanity has not been able to destroy the entire life-support system over substantial portions of the surface of Earth in a short space of time. Now, however, synthetic herbicides, insecticides, weather modification technology, and nuclear weapons have given people a greatly increased capability of committing ecocide.

If the widespread use of biocides continues, despite growing knowledge of the probable consequences, civilization may destroy much of the carrying capacity of this planet for human life. If that occurs, however, survivors will be able to explain what happened as a result of greed and stupidity — after all, no one really wanted the unhappy side effects of synthetic biocides.

But no such excuse can be made for the practice of ecocide by the armed forces of the United States in the Vietnam war. In that war the United States attempted to win by substituting firepower for manpower. It managed to retain control of the main cities, so the vast majority of its firepower was concentrated on the countryside, rather than on population centers. The people in rural areas and agricultural and natural ecosystems bore the brunt of the assault both with "conventional" weapons and with herbicides and other technologies for ecosystem alteration and destruction.

According to a report of the National Academy of Science (NAS), through 1971 — when spraying stopped — some 1.5 million hectares of South Vietnam, 10 per cent of the country had been treated with about 20 million gallons of herbicides, applied at an average of 13 times the dose recommended by the U.S. Department of Agriculture for domestic use. Most of the spraying was done in forests in an unsuccessful attempt to expose the enemy and secure the area. At such dosages some of the trees, especially mangroves, can be not just defoliated but killed by a single application. Multiple applications kill other kinds of trees. The American Association for the Advancement of Science Herbicide Assessment Commission (AAAS-HAC) reported that perhaps half the trees in the mature hardwood forests north of Saigon were damaged.

One of the herbicides widely used in Vietnam was picloram, which is both potent and (like DDT) persistent. It is so dangerous that the Department of Agriculture has not licensed it for use in the cultivation of any American crop.

By the end of 1971 at least 100,000 hectares of cropland in South Vietnam had been sprayed with herbicides with the stated purpose of denying food to enemy soldiers. This was done in an attempt also to starve civilian populations sympathetic to the National Liberation Front and to force people to move from the countryside into cities, where they would be easier to control.

As our military was well aware, the first to suffer when food becomes scarce in wartime are not soldiers, but children, old people, and pregnant and nursing women. The AAAS-HAC reported that virtually all the crops destroyed would have supplied food to civilians, especially to the Montagnard tribesmen of the central highlands.

The general effect of the herbicide programs was to degrade Vietnamese ecosystems. Forests attacked with herbicides have been invaded by bamboo, and some areas may have undergone laterization. The danger of laterization is especially severe in farmland attacked with herbicides.

Large areas of mangrove forest, which grows along waterways and plays a critical role in the maintenance of economically important fisheries, have also been destroyed. According to the NAS report, about 36 per cent of the mangroves in South Vietnam (0.6 per cent of the country's area) have been killed, and the forests will not recover for perhaps a century.

The long-term consequences of the programs are incalculable, since they have also certainly had a profound effect on the fauna of the area (and animals, in turn, are important to the plants for pollination and seed dispersal) and probably also on the organisms of the soil, which are essential to its fertility.

Very often in Vietnam plant succession in repeatedly defoliated areas has resulted in the establishment of an impoverished community dominated by bamboo or low grasses of no use to the Vietnamese.

Herbiciding operations also may have left the Vietnamese with a legacy of birth defects, as some of the constituents of herbicides are teratogenic.

Due in large part to the AAAS-HAC report and other pressures from the American scientific community, the use of herbicides as a military weapon in Vietnam was discontinued in May of 1971.

Herbicides were not the only ecocidal weapon used in Indochina. In 1967/ 1968 alone, more than 3.5 million 500-to-750-lb (225-340 kg) bombs were dropped on Vietnam, each creating a crater as much as 14 meters across and 9 meters deep.

If those craters were placed in a line, they would stretch for some 50,000 kilometers, a distance longer than the circumference of Earth. They occupy 40,000 hectares.

From 1965 to 1973 the U.S. military detonated 7 million metric tons of high explosives in South Vietnam, the equivalent to the explosive power of 400 Hiroshima-size atomic bombs-almost half a ton of munitions for each hectare of South Vietnam's land area.

The area of craters created in South Vietnam by bombs and shells was some 150,000 hectares 2 billion cubic meters of soil were displaced, and millions of additional hectares were contaminated with missile fragments. In many areas of Vietnam, peasants are afraid to reoccupy bombed fields because of the danger from unexploded ordnance.

QUANTITATIVE POLLUTANTS WITH ECOSYSTEMIC EFFECTS

Most attention to the impacts of pollutants on natural ecological systems has been directed at pesticides. And, with the exception of radiological pollutants of the effects of a given pollutant on an entire terrestrial ecosystem have been confined very largely to synthetic insecticides. It is clear that many other qualitative pollutants (such as the PCBs) must have ecosystemic effects but they have at most been guessed at.

We now turn to a consideration of some of the quantitative pollutants — petroleum, mercury, nitrates, phosphates, and radiation. Ecological systems have had to deal with all of them for eons, but now they may be present in quantities that constitute overloads for the systems in which they occur.

Petroleum Pollution

One ocean pollutant that has received a great deal of publicity — much of it the result of spectacular supertanker accidents — is oil. In 1970, when Thor Heyerdahl sailed a papyrus raft across the Atlantic Ocean, he reported, "Clots of oil are polluting the mid-stream current of the Atlantic Ocean, from horizon to horizon." Oil has reached the sea in a variety of ways.

Among them are the massive accidental spills from tankers that have received so much publicity, as well as many smaller spills. Those probably account for less than 0.1 per cent of the total amount of oil transported at sea. But the volume is so huge (about 1.2 billion metric tons per year, or 60 per cent of all sea-transported goods) that the spills and oil purposely flushed from tankers amount to a considerable influx.

Another source is fuel oil spilled from ships not involved in transporting oil, especially in connection with their refueling operations. In addition to shipping spills, there have been accidents in the extraction of oil from the seafloor, of which the 1969 Santa Barbara leak is the most widely known example. Some oil reaches the sea in sewage as wastes from automobile crankcases and from industry (including refineries).

There is also the natural influx of oil into the oceans from submarine seeps. Those may be important locally, but on a global scale the input from seeps appears to be about 10 times smaller than human contributions. Estimates of the magnitudes of major sources of petroleum in the oceans.

Table. Sources of Petroleum in the Ocean

Source	Flow (million MT/yr)
Marine transportation	2.1
Coastal oil refining	0.2
Offshore oil production	0.1
Industrial wastes	0.3
Municipal waste & urban runoff	0.6
Atmospheric rainout	0.6
Natural seeps	0.1-0.6

The effects of oil pollution on oceanic ecosystems, beyond the immediate and obvious decimation of populations of fishes, shellfishes, and seabirds at the site of a spill, are now being discovered, thanks in part to a relatively small spill that occurred in 1969 near the Woods Hole Oceanographic Institute in Massachusetts. Effects vary with the type of oil, how far from shore it is spilled, and how long it can "weather" before reaching shore, and what organisms live in the affected areas.

Some components of oil are toxic; others are known carcinogens. Weathering can reduce their toxicity, but the carcinogenic components are long-lasting. Oil washed to shore lingers on rocks and sand for months or years, but the marine life may need a decade or more to recover, even after the oil is no longer obvious. Populations of some seabirds, especially the diving birds, which are most susceptible, have been drastically reduced in recent years, probably largely because of oil contamination. Besides killing thousands of birds directly, oil toxins have been shown experimentally to affect their reproduction and to reduce the viability of their eggs when applied directly to them. Other effects are more subtle but nonetheless cause for concern — such as the altering of feeding behaviour in lobsters.

Detergents used to clean up oil spills have been found to make the situation worse in many cases. Not only are the detergents themselves toxic to many forms of life but they disperse the oil and spread it into new areas. They may also render it more easily absorbable by small marine organisms by breaking it up into droplets.

Immediately after the Massachusetts spill, there was a 95-per cent mortality of fish, shellfish, worms, and other sea animals in the immediate area. Nine months later, repopulation had still not taken place. Surviving mussels failed to reproduce. Some constituents of the oil were still present and killing bottom-dwelling organisms eight months later.

Surviving shellfish and oysters took in enough oil to be inedible and retained it even months after having been transplanted at a distance. According to marine biologist Max Blumer, there is a possibility that the carcinogenic constituents of oil, which are absorbed unchanged by those and other small organisms, may in time be incorporated into and contaminate entire food chains.

Although large amounts would probably spoil the taste and smell of conventional foods, toxic and carcinogenic effects are certainly possible at levels of contamination too low to taste or smell. It has also been suggested that higher levels of contamination might go undetected in fish-protein concentrates, which some people consider to be a potential future source of protein for human beings.

To the extent, however, that the carcinogens in oil would accumulate mostly in the fat of fishes rather than in muscle, the carcinogenicity problem in fish-protein concentrates would be minimized.

Some effort is being made to develop bacterial strains that can quickly degrade spilled oil; more efficient technical means of collecting it before it reaches shore are also being developed. In view of the amount of oil already polluting the oceans and the increasing potential for further spills represented by the rising volume of oil being transported by sea and the increasing size of tankers, efforts to deal with such accidents should be vigorously pursued.

At the same time, much can be done in the area of prevention without further research. Oil can and should be removed from sewage; safety regulations for both drilling platforms and tankers can be tightened and enforced; the flushing of tankers can be prohibited; and other ships can be forbidden to waste or discard oil products of any kind. Legislation covering some of these points has been enacted in the United States, but, with the exception of tracing one tanker responsible for a spill and prosecuting its captain, enforcement before the end of 1976 was virtually nil.

Mercury and Other Heavy Metals

Mercury, in organic forms such as methyl mercury, has been found experimentally to reduce photosynthesis in marine and freshwater phytoplankton. Photosynthesis was significantly inhibited at concentrations in water of 0.1 parts per billion. When concentrations of mercury reached 50 parts per billion, growth stopped almost completely. Mercury's effects on phytoplankton resemble those of DDT; like the chlorinated hydrocarbons, it also tends to be concentrated in food chains.

Thus, its *potential* impact on oceanic food webs may be serious and similar to that of DDT. The key question in both is whether environmental concentrations of these chemicals, singly or in combination, will ever be high enough to have a significant effect. Like DDT, enough inorganic mercury has already accumulated on lake- and streambeds in North America to provide a serious threat (if the mercury is converted to the soluble organic form) to freshwater and estuarine life for decades to come, unless a way is found to remove it.

San Francisco Bay, alone, was estimated by the U.S. Geological Survey in 1971 to have some 52 metric tons of mercury on its bottom, with concentrations ranging from 0.25 to 6.4 parts per *million* in bottom sediment.

What effect mercury has high in oceanic food chains is not clear, however. High levels of mercury in oceanic fishes such as tuna and swordfish may well arise through food-chain amplification of natural oceanic mercury levels, independent of pollution contributions, but that does not necessarily mean that ecological effects from anthropogenic increments of mercury will be absent. It is certainly possible that mercury could have deleterious effects on terrestrial food chains.

It has been shown, for instance, that mercuric chloride can be responsible for eggshell-thinning in birds; but high concentrations would appear to be necessary to elicit that effect, and it has been demonstrated *only* for mercuric chloride, not methyl mercury, which is the predominant form in the environment.

There is evidence that mercury may influence community structure by affecting plants and insects as well as higher vertebrates. Laboratory experiments have been done using tomatoes, corn, and beans grown in hydroponic solution with aphids feeding on the plants and lacewings feeding on the aphids. Concentrations as low as 0.006 parts per million (ppm) of methyl mercury hydroxide in the hydroponic solution led to reductions in biomass of the tomatoes, while corn and beans were about a half and a tenth as sensitive respectively.

The aphids concentrated mercury to a level about 1700 times the concentration in treatment solutions and suffered a significant decrease in fecundity and lengthened development time when the concentration in the treatment solution was 0.02 ppm. The lacewings concentrated mercury to 4200 times that in the hydroponic solution.

The long-term ecological effects in terrestrial food chains and in the seas of other heavy metals (those with density 5 times that of water or more), such as lead, cadmium, chromium, nickel, and thallium, are not known. But, since they are known to be toxic to many forms of life, it would be unreasonable to assume their effects to be negligible. Some, for example, can be potent inhibitors of photosynthesis.

Air Pollution and Ecosystems

Air pollution assaults humanity indirectly, as well as directly. The ecological systems of Earth clearly are degraded by the quantitative and qualitative pollutants that are injected into the atmosphere, although we are just beginning to understand some of their effects in detail. Many others are unknown or, at best, guessed at.

In some cases damage is dramatically obvious. For instance, in northern Ontario, Canada, sulfur dioxide pollution has virtually destroyed the natural forest ecosystem for 8 kilometers downwind of an iron-sintering plant and has severely damaged it 16 kilometers or more away. Damage can be detected from the air 30 kilometers from the site.

Other damage, less obvious but more important because of the area involved, is done by acid rains falling over a large portion of Europe and the northeastern United States. Sulfur oxides and nitrogen oxides are, as described earlier, injected into the atmosphere in large quantities by the burning of fossil fuels. Those oxides change in chemical reactions in the atmosphere into strong acids—sulfuric and nitric, respectively. The overall effects of acid rains on ecosystems are not well understood, but they have the potential for serious damage.

The rains may modify rates at which nutrients are leached from soils and foliage; they can reduce or exterminate fish populations by acidifying streams; they can affect soil microorganisms, especially those responsible for nitrogen fixation; they can influence the behaviour of bacterial and fungal pathogens; they can enhance uptake of the toxic heavy metal cadmium from the soil; and they may interact synergistically with herbicides and other synthetic compounds.

Acid rains were one of the first environmental phenomena to attract attention because of their international implications. Sweden made them the topic of a case study for a United Nations Conference on the Human Environment in Stockholm in 1972.

Sweden's forests have suffered a reduction in growth rates starting in the 1950s, thought to be due to acid rains, and salmon and trout have disappeared from Swedish streams and lakes where the pH has fallen much below 5. The sulfur that rains down on Sweden apparently originates in northwest Europe, especially in the United Kingdom. (Besides the potentially serious ecosystemic effects, Sweden was concerned about the corrosion of structural materials by the

Air pollution can have direct deleterious effects on ecosystems, in addition to causing acid rains. Ozone, which plays such a crucial and positive ecological role in the stratosphere, is extremely damaging to plants. It enters leaves through the pores (stomata) used for normal gas exchange and is thought to alter the permeability of the membranes of cells there.

This causes nutrient and electrolyte imbalances and the cells die. Chronic exposure to ozone may weaken plants and make them more susceptible to disease, or it may age them prematurely (cutting yields in crops like citrus)—all without outward signs of injury. Sensitivity in plants is such that a year or two of exposure to relatively low concentrations [15 parts per hundred million (pphm) can significantly alter the composition of a community—with results that could be deleterious].

Ozone is an important component of photochemical smog, which contains other oxidants damaging to plants—peroxyacetyl nitrate (PAN) and nitrogen dioxide. PAN is very phytotoxic, more so than ozone, but ozone normally occurs in much greater concentrations.

Photochemical smog originating in Los Angeles (with ozone concentrations of as much as 70 pphm) has had a dramatic impact on the ecosystems of southern California. Some 125 kilometers east of the city two-thirds of the 65,000-hectare San Bernardino National Forest has been affected, and within the affected area 50 per cent of the trees have been moderately or severely damaged.

In the northeastern United States there is also extensive forest damage from photochemical smog, and concern over ozone damage is growing in the United Kingdom. In those areas the ozone concentration is only one-third or less that of the Los Angeles basin, but ozone is more damaging to the plants in those more humid areas.

The effects of air pollutants on plants can be both complex and subtle. Sulfur dioxide damages plants directly (and differentially) in addition to contributing to acid rain. In the United Kingdom in the summer, however, there is a problem with sulfur-deficiency on nitrate-rich fertilized farmland.

The sulfur-dioxide pollution helps to overcome that deficiency, thus stimulating plant growth. There is evidence, however, that this benefit is more than balanced by deleterious effects of the sulfur dioxide on winter growth. Much investigation needs to be done to elucidate such interactions and to determine what effects pollutants have on mutation rates and on the production and germination of seeds.

It is also important to identify the ultimate impact of air pollution not just on the plants but on all components of an ecosystem. For instance, fluoride is directly injurious to plants and is also concentrated by them. Under conditions of severe pollution, forage can be so contaminated as to produce fatal fluorosis in cattle, but the effect of lower levels of fluoride pollution on herbivores is unknown. It seems likely that such fluoride pollution would have local ecosystemic effects at most.

NITROGEN, PHOSPHORUS, AND EUTROPHICATION

Nitrogen and phosphorus are nutrients required in large quantities for the growth of terrestrial and aquatic plants. (The natural cycles of both elements and the magnitudes of some of the perturbations imposed by civilization on those cycles.) The heavy use of nitrogen and phosphorus fertilizers has arisen because the availability of the two elements to plants is often the factor that determines how much plant material can be grown.

That is, if sunlight, water, and carbon dioxide are available in abundance, adding nitrogen or phosphorus (whichever is limiting in the given situation), or both, can increase the yield. This situation becomes a liability when extra phosphorus and nitrogen are made available inadvertently to ecosystems in which extra plant growth is undesirable.

The most important example of that phenomenon is in aquatic ecosystems—rivers, lakes, and coastal waters. There, over fertilization can produce overgrowth of aquatic weeds that interfere with recreational uses of the affected bodies of water. More seriously, over fertilization can lead to rapid growth of various algae, which can have several undesirable consequences: reduced transparency of the water, cutting off bottom plants from sunlight; formation of surface scums that are unsightly, smell bad, and discourage recreation; the generation of unappealing taste and odor in drinking water; the creation of a concentration of dissolved organic material conducive to the support of various disease organisms that would not prosper in less rich waters; the alteration of the *relative* abundance of the various algal species present, with adverse effects on food chains established under the previous conditions; and the depletion of dissolved oxygen, with resulting loss of fish life when the algal over growths or blooms die off and decompose.

How can oxygen-depletion arise from the overgrowth of plants, inasmuch as they must have produced as much oxygen in their lifetimes as is needed for their decomposition? The answer is twofold. First, oxygen is continuously exchanged between water bodies and the atmosphere. Second, dead algae often sink and, as they decompose, deplete the oxygen content of deeper layers of water where the oxygen content already tends to be lower than that of the surface layer.

The oxygen depletion is aggravated in many temperate-zone lakes by a phenomenon called *thermal stratification*. There is a warm layer of water at the surface in summer called the epilimnion, which is well mixed by the action of wind and waves and in which the temperature is almost uniform. Below the epilimnion is a layer called the thermocline, in which the temperature decreases 1°C or more for every meter of depth.

The lowest layer contains the coldest water, is almost uniform in temperature, and is called the hypolimnion. There is almost no vertical mixing in the thermocline and hypolimnion. The stratification is maintained because the density of water varies with temperature and the maximum density occurs at 4°C. Thus, as long as the coldest water in the lake is not less tnan 4°C, the coldest water will be on the bottom, slightly less cold (and hence less dense) water just above it, and so on.

A temperate-zone lake that is stratified this way gets thoroughly mixed vertically twice a year. In the autumn, surface temperatures start to drop, eventually falling below the temperature of the deeper layers. Having thus become denser than the underlying layers, the surface water sinks. In freezing climates, the turnover process continues until the thoroughly mixed lake has been uniformly cooled to 4°C; as the surface water cools further, it no longer sinks because it then becomes less dense than the 4°C water below.

The eventual result is a winter stratification with ice on top and water slightly above freezing below. Then, in the spring, the warming of the surface layer to the density maximum at 4°C causes that water to sink into the slightly cooler and less dense water below, and the lake turns over again.

Summer thermal stratification keeps oxygen-rich water at the surface from mixing into deeper layers; sustaining animal life in the deep water therefore depends on whether the supply of oxygen provided by turnover in the spring lasts until the next turnover in the autumn. If a lake is clear enough, some light penetrates to the hypolimnion so that the oxygen provided by the turnovers may be supplemented by photosynthesis there.

Such oligotrophic lakes are clear, deep, and relatively low in plant life. Maintaining this condition depends on a shortage of nutrients. In eutrophic lakes, which have an abundance of nutrients, algal growth cuts off sunlight near the surface, and the decomposition of the dead organic matter raining down from above depletes oxygen in the hypolimnion, which can be replaced neither by photosynthesis (because there is no light) nor by mixing (because of stratification). Eutrophic lakes therefore cannot support populations of those species of fishes that require cold water, because in summer the deep, cold water contains no oxygen.

Natural processes cause all lakes to undergo the transition from oligotrophic to eutrophic over a long period of time. Specifically, the transition is brought about by the flow of nutrients in streams that drain the surrounding area, with the effect not only of enriching the organic content of the lake water but also of filling the lake gradually with dead organic matter that settles on the bottom. This natural transition is called eutrophication.

The same term has been given to the acceleration of the process that takes place when the activities of civilization lead to substantial increases in the flow of nutrients into a lake. In everyday usage the term has become synonymous with *overfertilization* of any body of water—lake, river, estuary, bay, or near-shore ocean waters. The specialist's term for overfertilization of a body of water due to human activity is *cultural eutrophication.*

Several anthropogenic pathways to water exist for both nitrogen and phosphorus. Both substances are applied to fields in inorganic form in fertilizer, and both are present in sewage and in the wastes from livestock feedlots. Nitrates originating in vehicle exhausts fall in the rain, and phosphates added to detergents to improve their "cleaning power" become ingredients of municipal waste water. Neither nitrates nor phosphates are removed effectively from sewage by primary or secondary treatment plants.

Phosphorus is the limiting nutrient in the growth of aquatic plants in far more circumstances than nitrogen is. One reason for this is that some aquatic plants (including blue-green algae) can convert atmospheric nitrogen that has become dissolved in the water into usable chemical form. The growth of those plants is unlikely to be limited by lack of nitrogen.

The other major reason is that nitrogen compounds are in general much more soluble than phosphorus compounds, so the former migrate much more readily through soil and into waterways. The higher mobility of nitrogen compounds leads to much greater concentrations of nitrogen than of phosphorus in most natural bodies of water—enough so that phosphorus is the limiting nutrient for plant growth.

Because phosphorus compounds are so tightly bound to the soil, even very heavy fertilization does not lead to the leaching of much phosphorus into groundwater and surface water. Fertilizer phosphorus that reaches waterways does so almost entirely by being carried along in eroded soil particles.

Even in the water, most of that phosphorus remains in suspension (not in solution) so it is not readily accessible for use by aquatic plants. By far the largest source of soluble phosphorus in waterways is municipal sewage, which contains phosphorus both from excrement and from detergents.

Many examples of the cultural eutrophication of lakes, due mainly to phosphorus additions in sewage, have been documented. Some of them, such as Lake Washington in Seattle, some of the lakes around Madison, Wisconsin, and Lake Onondaga in New York, have made dramatic recoveries when citizen action has stopped the sewage discharges or restricted the use of phosphate detergents.

The most famous example of advanced cultural eutrophication, Lake Erie, has been less fortunate. The streams that feed Lake Erie drain a large area that is heavily populated, highly industrialized, and intensively cultivated. Controlling the sources of phosphorus over such an area would be an enormous undertaking.

The lake itself is relatively shallow (66 meters maximum depth, compared to between 230 and 406 meters in the other Great Lakes), and a thick layer of muck on the bottom contains vast quantities of nutrients that under appropriate conditions might redissolve into the lake. Large algal blooms are so common in Lake Erie, that only a few days after the onset of thermal stratification in the summer the oxygen in the hypolimnion is essentially gone.

The relative abundance of fishes in the lake has undergone a drastic transition over the past fifty years, characterized by replacement of species highly valued by people as food (lake trout, walleye, blue pike) with species considered less desirable (alewife, carp, sheepshead). Some of the changes in fish populations have been brought about not by cultural eutrophication alone, of course, but by the combined effects of eutrophication, the introduction by people of the sea lamprey and the alewife, and the heavy pollution of the lake with pesticides, industrial chemicals, and heavy metals.

Lake Erie is not dead, but it clearly is suffering from premature old age combined with a case of chronic poisoning. Unless it somehow proves possible to prevent permanently the resuspension of the inventory of toxins and nutrients in Lake Erie's bottom sediments, its condition—unlike those of smaller lakes—will not be reversible.

Although the best known and probably most serious examples of cultural eutrophication are lakes, rivers and estuaries are not immune to the problem. In a major ten-year survey of the twenty-two most important rivers and harbors in the United States (ranked by length and volume for rivers and by size of adjacent populations for harbors), 84 per cent of the sections sampled exceeded phosphorus and phosphate concentrations associated with potential eutrophication, and 54 per cent showed an increase in those concentrations between the period from 1963 to 1967 and the period from 1968 to 1972.

Nitrate concentrations exceeded those associated with potential eutrophication in 25 per cent of the sections sampled and had increased between the two periods in 74 per cent of the sections. Contrary to a fairly widespread impression, the circulation patterns in typical estuaries, and even those along open ocean coastlines, are not conducive to rapid mixing of pollutants into the deeper water offshore; therefore, cultural eutrophication of those ecologically crucial regions is not a problem that can be ignored safely.

Even though phosphorus rather than nitrogen is most often the governing nutrient in cultural eutrophication, overuse of inorganic nitrogen fertilizers may prove to be the greater environmental hazard for other reasons. One reason is that runoff of nitrates into surface- and groundwater supplies can lead to concentrations great enough to pose a direct toxic danger to the health of animals and human beings, both as methemoglobinemia and as a potential carcinogen.

Secondly, the accelerated release of nitrous oxide (N_2O) into the atmosphere may in time adversely affect the ozone layer with implications for human health and future food production. A further reason for concern about the prevalent pattern of using nitrogen fertilizer is the effect of those practices on soil humus, which is essential to the natural nitrogen cycle.

Humus, the organic matter in soil, is a poorly understood complex of compounds of high molecular weight. In natural soils inorganic nitrogen normally constitutes less than 2 per cent of the nitrogen present. Most of the nitrogen is tied up in the large organic molecules of humus, which are derived from such varied sources as the fibrous remains of woody plant tissues, insect skeletons, and animal manure.

The presence of humus makes the soil a favorable medium for the complicated chemical reactions and mineral transport needed for the growth of higher plants. Bacteria in the soil decompose humus to form nitrates and various other nutrient substances required by plant roots.

Roots require oxygen in order to take up nitrates and other nutrients, but oxygen is not available if the soil is tightly compacted, nor is water absorbed well. Thus one important function of humus is to maintain the porosity of soil, which permits oxygen to penetrate to the roots of plants and promotes the retention of water. Greater capacity to absorb and hold water reduces runoff and soil erosion after heavy rain and prolongs the retention of soil moisture during times of drought. On several counts, then, humus is essential for the healthy growth of plants, and particularly (as experienced farmers know) for good crops.

In natural soil systems the nitrogen cycle is tight; not much nitrogen escapes to the atmosphere or is removed from the soil by leaching or surface runoff. It has been shown experimentally (as well as by centuries of farming experience) that by maintaining the supply of humus the fertility of soil can be perpetuated.

This is not possible when fertilizers containing inorganic nitrogen are employed, unless organic carbon (in such forms as crop residues, sawdust, or straw) is supplied to the microorganisms in the soil. The undesirable decline of humus, which often occurs under heavy inorganic fertilization, is due to the failure of farmers to return crop residues (and thus carbon) to the fields.

The decline is not caused by any deficiency in the fertilizers themselves. Indeed, if carbon is supplied in the proper proportion to inorganic nitrogen, the supply of humus can be increased and the quality of the soil improved.

If attempts are made to maintain soil fertility by continued applications of inorganic nitrogen fertilizers alone, the capacity of soil to retain nitrogen is reduced as its humus content drops. In humus, nitrogen is combined into nonsoluble forms that are not leached from the soil by rainwater. Depletion of humus "loosens" the soil cycles and permits large amounts of nitrate to be flushed readily out of the soil. Only about 50 to 60 per cent of the applied nitrogen is incorporated into the crops. This loss plus depletion of the original humus leads to the need for more and more fertilizer, in a vicious circle.

The serious consequences of such human interventions in the nitrogen and phosphorus cycles call for strong remedial measures. The only successful steps against phosphorus pollution that had been taken in the United States as of 1975 were restrictions on the phosphorus content of detergents, a few local bans on the sale of phosphorus-containing detergents, and diversion of sewage and industrial wastes from a number of acutely suffering lakes.

With specific regard to nitrogen, there are only restrictions on the nitrate levels in foods. Much more systematic measures are required, the most important of which would be closing the open nitrogen and phosphorus cycles of agriculture by returning the organic "wastes"—now contributing to the pollution problem—to the soil.

Human excrement, for instance, is considerably richer in both nitrogen and phosphorus than cattle manure. In the People's Republic of China, human waste is carefully collected and used largely untreated, along with manure and agricultural wastes. These organic fertilizers still comprised about half of China's total fertilizer applications in the early 1970s.

The direct use of human (or livestock) waste as fertilizer presents opportunities for the transmission of many diseases and parasites. Some methods of treatment, such as aerobic or anaerobic digestion by bacteria, however, appear to prevent this, at least for most fertilizing purposes. Treated sewage water and sludge (the concentrated residue that settles out of sewage water) is used successfully as fertilizer in many parts of the world, including the United Kingdom, Australia, Fed- eral Republic of Germany, and a few places in the United States. At least one operating plant in the United Kingdom, run by the local water authority, sells recovered water to industry and treated liquid and dried sludge fertilizer to local farmers.

It uses the sludge gas produced by the digestion process to generate electricity that runs the plant, a nearby sewage pumping station, and an incinerator, and cooks meals for a group of senior citizens. "Waste" heat from the generators is used for the digesters. Although the costs of collecting, treating, and distributing sewage have been a major obstacle to using it as fertilizer, this system apparently costs little more than conventional sewage-disposal methods.

In the United States, the manure of cattle produced on feedlots alone far exceeds in volume that produced by the human population, even though phosphorus and nitrogen concentrations are lower. The nutrients in that material, as well as manure from similarly managed chickens and pigs, also could and should be returned to the soil. Some 80 to 90 per cent of these animals are in feedlots for part or all their lives. In the past, huge volumes of feedlot sewage were discharged into rivers, causing severe pollution.

The EPA has made reduction of pollution from that source to zero by the mid- 1980s a goal, which should provide incentive for alternative uses. If all feedlot sewage in the United States were used for fertilizer, it would provide more than enough for all U.S. acreage planted in corn. If augmented with municipal sewage, garbage, and other wastes, it would, of course, go even farther.

Other potential sources of organic fertilizing nutrients include agricultural wastes (some of which can also be used as livestock fodder), treated garbage, and food processing wastes. Most of these presently are wasted: they are discharged in waterways, put in landfill, or burned in incinerators.

The values of these materials as fertilizer must, of course, be balanced against their values for other uses, such as supplementary livestock feed and energy sources, taking into consideration the energy and other costs of the alternatives. But it is clear that merely throwing them away and then cleaning up the mess (both costly and energy-consuming exercises) is the east intelligent solution.

The most promising method of recovering and recycling the nutrients from manure, human sewage, crop residues, and other organic wastes—while deriving energy as well—is with biogasification plants like the one described above.

Such facilities, which can range in size from backyard digesters to large factories, have the dual advantage of producing needed fuel from the hydrocarbon content of the wastes and leaving the nitrogen, phosphorus, and other nutrients behind in a residue much more compact than the original wastes. Thus, transportation costs and handling problems associated with using those residues as fertilizer should be less than those of using the uncompacted raw material.

The prospects that this approach can compete economically with using inorganic fertilizers are improved by the rising cost of the latter, particularly the very energy-intensive inorganic nitrogen fertilizer whose cost is closely tied to the cost of energy. Even if fertilizer from organic wastes should turn out to be more expensive at first than inorganic fertilizers, however, its use should be strongly encouraged by regulations and subsidies, in recognition of its ecological advantages.

The U. S. National Aeronautics and Space Administration (NASA) has recently been conducting experiments using water hyacinths to purify waste water. These aquatic plants, generally considered a pest in warm regions around the world, have proven remarkably efficient at absorbing nutrients and other pollutants from sewage water. The dried plants then can be used to make biogas and/or for fertilizer or livestock feed.

Naturally, care must be taken to control the new problems that innovative approaches to old problems often bring. (When phosphorus was discredited on ecological grounds as an ingredient in detergents, for example, one of the first substitutes that was tried was NTA—nitrilotriacetic acid. Then evidence appeared that compound was a potent mutagen, and its use was discontinued.)

The precise effects on soil organisms of fertilizing with various kinds of manure are not yet known, although they are being investigated. Results of comparing yields under different fertilizing regimes—all organic, all inorganic, or a combination—in small, controlled studies indicate that the mixed approach may be best for many crops.

Perhaps the most serious potential problem associated with continuously recycling nutrients by means of biogasification and application of the residues as fertilizer is accumulation of toxic heavy metals, whose concentration could increase at each cycle. While trace amounts of many of these—copper, cadmium, cobalt, for instance-are essential to life, high concentrations accumulated in soil and concentrated by crops can be deadly.

Some experimental schemes for processing urban sludge are attempting to remove these and other toxic substances to be recycled in industry. Better yet, the industries themselves should be encouraged to remove such toxins before discharging them into waste water. In LDCs, most people still live in rural areas, which would minimize both the problem of transporting their wastes to farms and the likelihood of significant pollution with heavy metals or other toxic materials.

Thermal Pollution

The term *thermal pollution* is most commonly used to describe the discharge of waste heat from thermal generation of electric power or from industrial processes into natural bodies of water.

The ultimate fate of this energy—and of all other energy mobilized by human activity—is to find its way into the atmosphere and from there into space. The atmospheric aspects of this more general "heat problem" are considered later; here we consider only the effects of heat deposited directly into aquatic ecosystems.

Aquatic thermal pollution has been the focus of much research and considerable controversy. Some of the controversy has arisen because the effects are so different at different sites. Even very large power plants can discharge their heat to the ocean with minimal effect where deep water lies close to shore and where the local surface and underwater topography permit good mixing of heated effluent with colder water.

Undoubtedly, there are even circumstances where the heated effluent can be beneficial—for example, in accelerating the growth of shellfish.

Radiation and the Structure of Ecosystems

Ecosystemic effects of radiation have been investigated through the use of chronic irradiation as a perturbing influence in experiments on natural ecosystems. The experiments employed powerful sources of radiation (9500 Ci cesium-137, 3000-4000 Ci cobalt-60) that exposed adjacent parts of an ecosystem to thousands of roentgens per day and areas progressively farther from the source to doses that declined roughly as the square of the distance.

The general results of such experiments in oak-pine forest, the lichen flora of that forest, and the herbaceous flora of an old field produced the following generalities. The structure of the community changed as upright plant forms were differentially damaged and prostrate forms were placed at a relative advantage.

Dominant forms replaced one another along the gradient of radiation dosage as they do along natural environmental gradients, diversity declined with increasing dosage, and usually (but not always) net production decreased with increased dosage.

These changes are quite similar to those produced by other gradients of environmental stress, such as increasing altitude on mountains, approach to the spray zone on seashores, decreasing rainfall as the edge of the desert nears, or nearness to an iron-sintering plant. The reasons for these patterns are not completely understood, but they seem to be related to changes in the production/respiration ratio in various segments of the community.

For instance, trees are limited in size because, as they grow, the proportion of bark, which contributes heavily to respiration, increases compared to that of photosynthetic leaves. Stresses tend to decrease photosynthesis rather than respiration; therefore they affect trees more than smaller plants, which do proportionately less respiring.

Structural changes in vegetation also change nutrient patterns and inevitably have a dramatic impact on the fauna. Like other pollutants, radioisotopes are concentrated by food chains—for instance, cesium-137 has been found to double in concentration with each step in Alaskan lichen'! caribou'! wolf (or Eskimo) food chains, and plutonium concentrations in algae of 1000 to 6000 times that of their environment have been measured.

In general, ecologists tend to feel that genetic disturbances from radiation exposure in nonhuman populations of both plants and animals are not serious; they assume that selection will compensate for them. Effects of radiation on animals are diverse, but some generalizations are possible.

For animal populations, the major stress seems to be effects on reproductive processes rather than shortened life spans. Long-lived species are more sensitive to radiation than short-lived ones; therefore, since predators are generally longer-lived than herbivores, they would be differentially removed from irradiated ecosystems. In this way, radiation effects may be very similar to those of pesticides.

It would appear that potential levels of radiation releases from nuclear-power enterprises, which are of great concern because of their direct effects on humanity, would probably have little serious effect on other components of ecosystems. On the other hand, in the aftermath of thermonuclear war, radiation could be one of the major stresses on Earth's ecological systems.

ATMOSPHERE AND CLIMATE

The complicated geophysical machinery that governs weather and climate more carefully this machinery has been studied, the more intricate and, in many instances, delicately balanced it has seemed to be. The complexity and possible fragility of atmospheric and related climatic phenomena mean that the possible avenues of significant human intervention are many.

They include changing the abundance of atmospheric constituents that govern the atmosphere's transparency to solar and terrestrial radiation (for example, by adding carbon dioxide, particles, water droplets and water vapour, and by destroying ozone), altering the characteristics of the interface between the atmosphere and land or sea (for example, by agriculture, urbanization, deforestation, desertification, and oil spills), and perturbing natural energy flows with the release of heat from industrial activities.

It cannot yet be said that atmospheric processes, weather, and climate are thoroughly understood, even exclusive of the growing perturbations caused by the activities of civilization. Not surprisingly, then, the actual magnitude and present and future consequences of such human perturbations cannot be accurately and unambiguously stated even as individual effects, much less in combination.

It is not even known whether all the important potential effects have been identified—not to mention quantified. The sudden emergence in 1974 and 1975 of the probable role of spray-can propellants as a threat to the ozone layer is indicative of the sorts of further surprises that may yet be in store.

Clearly, the fact that knowledge of the possibilities is incomplete is anything but grounds for complacency. Enough is already known to make plain that civilization has become a force large enough *in theory* to disrupt atmospheric processes and climate not only on regional but on hemispheric and even global scales.

That technological civilization has reached a level of activity that implies significant potential influence over global processes so crucial to well-being, without commensurate understanding or control, should appall any thoughtful person. In the following text we survey some of the more important mechanisms for human intervention in atmosphere and climate, beginning with the threats to the ozone layer that became really credible only in the mid-1970s.

Threats to the Ozone Layer

In the atmosphere at sea level, only about I molecule of 100 million is ozone (chemical formula O_3). Most of the small atmospheric pool of this compound is concentrated, however, in the part of the stratosphere lying between 15 and 40 kilometers altitude. In that region the molecular concentration of ozone reaches about 15 parts per million.

That is still only a trace, but it is enough to play a crucial role in the interaction of the atmosphere with incoming sunlight. In particular, atmospheric ozone is the only effective shield of Earth's surface against ultraviolet solar radiation in the range of wavelengths that are damaging to plants and to animals. Moreover, the absorption of this radiation in the stratosphere and consequent heat production there are responsible for the stratospheric temperature inversion that suppresses mixing of the lower atmosphere (troposphere) with the air at the higher altitudes. This fact, and the mere circumstance that the absorbed energy does not reach Earth's surface, certainly have some effect on world climate. The stratospheric pool of ozone is the result of a balance between continuous processes that produce and destroy this substance. (The size of the pool varies with latitude and with time in response to a variety of natural factors.) Production takes place when molecular oxygen (O_2) is split by ultraviolet solar radiation and the resulting oxygen atoms (O) attach themselves to other oxygen (O_2) molecules:

$$O + O_3 \longrightarrow O_2 + O_2$$

$$2O_3 \longrightarrow 3O_2.$$

Destruction takes place by means of several different reactions, in which the net results are either or

$$O_2 \longrightarrow O + O,$$

$$O + O_2 \longrightarrow O_3.$$

The destruction reactions proceed most rapidly in the presence of certain catalysts: the hydroxyl radical HO (which comes from water vapour in the stratosphere), nitric oxide (NO), and atomic chlorine (Cl). All of them are scarce in the natural stratosphere, although crucial to its chemistry.

Activities of civilization that change the stratospheric concentrations of these catalysts also change the rate at which ozone is destroyed, thereby altering the production-destruction equilibrium and possibly reducing the stratospheric pool of ozone.

SSTs and Nuclear War

The first specific threat to the ozone layer to be taken seriously by people in the technical community was the proposed operation of a fleet of supersonic transports (SSTs). An early concern was that the water vapour introduced into the stratosphere in the exhausts of those aircraft would cause a reduction in ozone by increasing the stratospheric concentration of HO.

This particular concern subsequently proved to be misplaced; the real hazard to ozone in SST exhausts is nitric oxide, which is a much more effective catalyst for the ozone-destruction reactions than HO. A paper published by chemist Harold Johnston of the University of California at Berkeley in 1971 indicated that the nitric oxide emitted by a fleet of 500 SSTs of the U.S. design could reduce the stratospheric ozone concentration by an average of 20 per cent, with local peak reductions up to 50 per cent.

This finding was strongly attacked by SST proponents, but a combination of further research and exhaustive reviews by panels of experts under the auspices of the National Academy of Sciences and the United States Department of Transportation completely substantiated Johnston's work. One of the reviews, by the Climatic Impact Assessment Programme (CIAP) of the Department of Transportation, produced a final report written so obscurely that some journalists misread it as a refutation of Johnston's results; and a few made fools of themselves in nationally syndicated columns decrying the "environmental hysteria" that had stopped the SST. Most of the confusion caused by the CIAP report was caused by the prominence it gave to the conclusion that a somewhat smaller fleet of the lower flying, smaller-engined Concorde SSTs would not significantly disrupt the ozone layer, while it buried the confirmation of Johnston's conclusion about 500 U.S. SSTs in the fine print.

Nuclear bombs offer an additional avenue by which significant amounts of nitrogen oxides can be added to the stratosphere. The high temperature in nuclear fireballs produces nitrogen oxides in abundance from atmospheric nitrogen and oxygen, and much of that material is deposited in the stratosphere.

Although there is some controversy about the issue, several investigators believe that ozone depletion by atmospheric bomb tests conducted in the early 1960s was great enough (around 5 per cent) to show up in the relatively crude historical data that are available for stratospheric ozone concentrations.

Although depletion of ozone may seem at first glance to be the least of humanity's worries if a large nuclear war ever took place, that is not really so obvious. In some circumstances, the global depletion of ozone might have a greater effect on noncombatant nations distant from the conflict than would radioactive fallout.

Chlorofluoro Carbons

It was known as early as 1973 that free chlorine in the stratosphere would be an even more efficient catalyst than nitric oxide for destruction of ozone, but since no source of free chlorine that could reach the stratosphere was known at that time, little concern was expressed about this phenomenon. Class of compounds variously known as chlorofluorocarbons, fluorocarbons, and Freons could be dissociated by wavelengths of sunlight present only in the stratosphere, to release free chlorine.

Those compounds have been used very widely as refrigerants and as propellants in some aerosol spray cans. The most common ones have the chemical formulas $CFCI_3$ and $CF_2 Cl_2$. They are essentially inert to chemical reaction and relatively insoluble in water. The result of these properties is that virtually the total amount put into the atmosphere remains there, chemically intact, until it reaches the stratosphere.

The amount of chlorofluorocarbons thought to have reached the stratosphere by late 1974 was only enough to have caused about a I per cent decrease in the ozone—a change too small to be detected by existing techniques. Production of those compounds has been growing rapidly, however; in this situation, a large fraction of the amount ever produced has been produced only recently and has not had time to diffuse up to the stratosphere.

Some calculations have indicated that ozone depletion due to catalytic destruction by chlorine from chlorofluoromethanes could reach 5 per cent by 1990, even if production of those compounds had been halted at the end of 1975, and around 15 per cent by the year 2000 if production continued to grow at 10 per cent per year.

There are many approximations in such calculations, and more precise analyses and measurements were underway as this text was being written. The available information gives ample cause for serious concern, however, a position underlined by a federal scientific task force that reviewed the issue in mid-1975.

Assertions by the industries producing chlorofluorocarbons that no action should be taken to curtail production until proof of harm is persuasive *to them,* are so irresponsible as to defy imagination.

Nitrogen Fertilizer

The man-made threats to ozone discussed so far—SSTs, nuclear bombs, and aerosol propellants and refrigerants—share the characteristic that human well-being does not depend in any fundamental sense on the use of the offending products.

SSTs are not economically necessary, harmless substitutes can be found for most applications of chlorofluoromethanes (modifications of the finger-operated spray pump familiar in window-cleaner bottles could handle most aerosol-can applications), and it should be plain that nuclear bombs are a product the world can do without. Thus, the future magnitude of those threats *could* be limited without serious social cost merely by the exercise of good sense—that is, by not producing or using them any longer.

An intrinsically more difficult problem is posed by the possibility that nitrogen fertilizer is the source of yet another threat to ozone. As Orville Freeman, former U.S. Secretary of Agriculture, has pointed out: "A world without fertilizer would be a world without the capacity to sustain a billion of our fellow human beings."

The danger is that nitrous oxide (N_2O) produced from nitrates and nitrites in soil or water by denitrifying bacteria is oxidized in the upper atmosphere to ozone-destroying nitric oxide. Applying more nitrogen fertilizer may add more nitrous oxide to the atmosphere, which means more nitric oxide in the stratosphere, which means less ozone.

The intrinsic difficulty is the apparent reliance of immediate prospects for increased food production on greater use of nitrogen fertilizer. This threat to the ozone is less certain at this writing than the others discussed here, because of uncertainties in the atmospheric chemistry of N_2O and in the magnitude of flows in the natural nitrogen cycle. Some models of the processes at work suggest that ozone depletion from this cause is likely to be small and gradual.

If the worst proves to be true, however, forecasted increases in the use of nitrogen fertilizer could produce a *15-per cent reduction* in stratospheric ozone by the year 2025. That prospect would pose an agonizing dilemma; unlike its relationships to SSTs, bombs, and spray cans, civilization is *really* hooked on nitrogen fertilizer.

At the very least it is another argument in favour of using nitrogen fertilizers so they are maximally effective and minimally wasted, and for relying as much as possible on natural fertilizers (from which relatively little nitrogen "leaks" to the atmosphere.)

Other Threats to Ozone

There may be other compounds as dangerous to the ozone layer as those already mentioned, or more so. Carbon tetrachloride (CCl_4), for one, is added to the lower atmosphere by processes both natural and human. There is evidence that human activity is the principal source, and that a significant quantity of the carbon tetrachloride reaches the stratosphere.

It is conceivable that chlorination of drinking water and of sewage are important sources that have some effect on ozone. Bromine is known to be an even more efficient catalyst for the destruction of ozone than chlorine; bromides are used in gasoline additives, agricultural chemicals, and flame-retardants. Pathways to the stratosphere for bromides have not yet been evaluated.

Alkali metals such as sodium and potassium might destroy ozone if added to the stratosphere by nuclear explosions over oceans or by intentional seeding as a chemical/climatological weapon. Many of these possibilities may prove to be false alarms, but, as in Russian roulette, only one "loaded" possibility may be too many.

Consequences of Ozone Depletion

The existing ozone layer screens out more than 99 per cent of incoming solar energy in the wavelength region between 0.23 and 0.32 microns; radiation of those wavelengths is harmful to most forms of terrestrial life, and the small fraction that gets through is known to cause sunburn, skin cancer, and eye damage in human beings, as well as various kinds of damage to other animals and plants. Most scientists agree that terrestrial life did not evolve until the ozone layer was established.

The special effectiveness of ultraviolet radiation in damaging living systems is due to the sensitivity of the genetic material DNA and of proteins to those wavelengths. DNA exhibits a resonance—the capacity to absorb radiant energy very efficiently—at wavelengths around 0.265 microns. Proteins typically have resonances in the range of 0.275 to 0.285 microns. The energy absorbed when radiation of those wavelengths strikes the DNA and protein molecules damages them, causing the many disruptive effects now known, and probably others as well.

Analytic and computer models of varying degrees of sophistication have been used to estimate what increase in the ultraviolet radiation reaching the surface would be associated with different degrees of depletion of the stratospheric ozone. Although the relation between ozone depletion and increased ultraviolet at the surface varies with latitude and other factors, a rough average is that a 5-per cent decrease in ozone produces a 10-per cent increase in radiation.

The best-studied effect of such an increase is an increase in the incidence of human skin cancers, the magnitude of which is estimated to be about equal to the percentage change in radiation. Thus a 5-per cent decrease in ozone and a 10-per cent increase in ultraviolet radiation would produce 20,000 to 60,000 additional skin cancers per year in the United States alone, and perhaps 5 times as many worldwide. (Incidence of skin cancer is higher among light-skinned people than among dark-skinned.)

About 5 per cent of the skin cancers would be fatal. A 30-per cent reduction in ozone would be expected to produce on the order of 1 million additional skin cancers per year worldwide, of which perhaps 50,000 would be fatal.

Although an increase in skin cancer is the best understood of the effects of ozone depletion, it is by no means the only and not necessarily even the most serious such effect. Laboratory tests have indicated that increased ultraviolet radiation impairs the growth of certain crop plants and adversely affects a wide variety of organisms from bacteria to vertebrates.

Phytoplankton, the primary producers at the base of oceanic food webs, may prove especially vulnerable if the light-receptors that trigger the protective dive response to excess radiation are not sufficiently sensitive to ultraviolet light. Furthermore, many insects "see" ultraviolet light and rely on it for navigation; interference with their pollination function is but one of several conceivable consequences for insects of an increase in ultraviolet intensity.

Acceleration of the mutation rate of microorganisms because of

DNA's sensitivity to ultraviolet could have adverse effects on human beings through disease, causing more frequent development of strains of microorganisms against which people would have no resistance and of strains resistant to antibiotics.

The consequences for ecosystems of differential susceptibility among different organisms to increased ultraviolet radiation cannot be predicted with any accuracy, but such differential susceptibility is known to exist and the possibility of its resulting in serious instability cannot be excluded.

Non-biological damage from increased ultraviolet radiation is likely to include faster deterioration of paints, fabrics, plastics, and other materials. The magnitude and economic cost of such effects is uncertain—and dependent on the amount of increased radiation—but it would surely be significant.

Probably least certain of all is the impact of ozone depletion on climate. The situation is extremely complicated: ozone absorbs and reradiates radiant energy in the infrared part of the spectrum of wavelengths as well as in the ultraviolet part; so do products of some of the reactions that consume ozone, and so do most of the catalysts for the ozone-destroying reactions and their precursors (chlorine is such a catalyst, the chlorofluorocarbons are its precursors).

One recent study based on a relatively simple model found it possible that continued growth of chlorofluorocarbon production could produce a significant global warming by the year 2000, independent of the effect on ozone, merely because of a greenhouse effect in which those compounds would block outgoing infrared radiation.

Changes in the radiation absorbing and -emitting properties of the atmosphere not only change the direct radiation flows at Earth's surface, however; they also change the vertical temperature distribution in the troposphere and the stratosphere, which in turn changes the rate of vertical circulation. All of this could influence climate in ways that as yet defy detailed prediction.

Surface Modification of Climate Change

Whereas the importance to climate of processes high in the stratosphere has been appreciated only relatively recently, the relevance to climate of the condition of Earth's surface is more obvious. The most important features are:

- Presence or absence of water.
- Reflectivity, or albedo.
- Ability to transfer water to the atmosphere.
- Capacity to store heat.

- Topography and texture.

These interrelated features determine how much of the incident radiant energy is captured, how it is apportioned between surface and atmosphere, how the surface interacts with the winds, and how the shoreline interacts with ocean currents.

Civilization can influence the surface conditions by changing or removing vegetation, by damming and diverting rivers to form reservoirs and irrigation projects, by "reclaiming" land from shallow seas and bays, by covering land with pavement and buildings, by spilling oil on water, snow, and ice, and by releasing previously stored energy as heat.

Index

A

B

C

D

E

F

G

H

I

L

M

N